SpringerBriefs in Economics

SpringerBriefs present concise summaries of cutting-edge research and practical applications across a wide spectrum of fields. Featuring compact volumes of 50 to 125 pages, the series covers a range of content from professional to academic. Typical topics might include:

- A timely report of state-of-the art analytical techniques
- A bridge between new research results, as published in journal articles, and a contextual literature review
- A snapshot of a hot or emerging topic
- An in-depth case study or clinical example
- A presentation of core concepts that students must understand in order to make independent contributions

SpringerBriefs in Economics showcase emerging theory, empirical research, and practical application in microeconomics, macroeconomics, economic policy, public finance, econometrics, regional science, and related fields, from a global author community.

Briefs are characterized by fast, global electronic dissemination, standard publishing contracts, standardized manuscript preparation and formatting guidelines, and expedited production schedules.

More information about this series at https://link.springer.com/bookseries/8876

Lily Kiminami · Shinichi Furuzawa ·
Akira Kiminami

Social Entrepreneurship, Social Business and the Multi-functionality of Urban Agriculture

Mixed Methods Research on Japan and China

Springer

Lily Kiminami
Faculty of Agriculture
Institute of Science and Technology
Niigata University
Niigata, Japan

Shinichi Furuzawa
Faculty of Agriculture
Institute of Science and Technology
Niigata University
Niigata, Japan

Akira Kiminami
Department of Agricultural and Resource
Economics, Faculty of Agriculture
The University of Tokyo
Tokyo, Japan

ISSN 2191-5504 ISSN 2191-5512 (electronic)
SpringerBriefs in Economics
ISBN 978-981-19-1761-5 ISBN 978-981-19-1762-2 (eBook)
https://doi.org/10.1007/978-981-19-1762-2

This Springer imprint is published by the registered company Springer Nature Singapore Pte Ltd.
The registered company address is: 152 Beach Road, #21-01/04 Gateway East, Singapore 189721, Singapore

Acknowledgments

This research was supported by JSPS KAKENHI Grant No. 17K07989 (Study on the creativity of cities through improving agriculture amenity: Comparative analysis between Japan and China) and No. 20K06256 (Analysis of social entrepreneurship in urban agriculture: Creating shared value through social business). We wish to express our gratitude to the support.

Contents

Abbreviations

BCP	Brain competition policy
BFP	Bifurcation point
CB	Community business
CC	Creative class
CEIBS	China Europe International Business School
CFI	Comparative fit index
CGs	Community gardens
CLDS	China Labor-Force Dynamic Survey
COVID-19	Coronavirus disease 2019
CSR	Corporate social responsibility
CSV	Creating shared value
DSEM	Dynamic structural equation modeling
EC	Electronic commerce
EFP	Equifinality points
EMBA	Executive master of business administration
FAO	Food and Agriculture Organization of the United Nations
GEM	Global entrepreneurship monitor
GRP	Gross regional product
HC	Human capital
MAFF	Ministry of Agriculture, Forestry and Fisheries
MFA	Multifunctionality of agriculture
MFUA	Multifunctionality of urban agriculture
MM	Mixed methods
NGOs	Non-Governmental Organizations
NPOs	Non-Profit Organizations
OECD	Organization for Economic Co-operation and Development
RMSEA	Root mean square error of approximation
SB	Social business
SC	Social capital
SDGs	Sustainable development goals
SE	Social entrepreneurship

SEM	Structural equation modeling
SI	Social institution
SID	Social institutional direction
SIG	Social institutional guidance
TC	Thinking creatively
TEM	Trajectory equifinality modeling
UCAs	Urbanization control areas
UNCTAD	United Nations Conference on Trade and Development
UPAs	Urbanization promotion areas

List of Figures

List of Tables

Chapter 1
Introduction

Abstract Japanese cities are facing the problems of modern society such as aging, environment, land, housing and transportation. On the other hand, in China, while the solution of the three agricultural problems and the construction of new rural areas are listed as important policy issues in the midst of rapid economic growth, the pressure of urban development is extremely strong. However, recent studies have reported that residents' creativity is one of the important factors for urban development. In other words, it is possible to increase the amenities of the city (the attractive factor of agriculture from the aspect of enjoying multi-functionality) by promoting urban agriculture to attract the people with high creativity and productivity. And those highly creative people have the potential to solve social issues in the region and contribute to the sustainable development of the city by engaging in social business. Therefore, the purpose of this study is to elucidate the mechanism that creative people with high level of social entrepreneurship solve regional issues through urban agriculture as a social business and contribute to sustainable urban development. It will mainly focus on the megacities, Tokyo (Japan) and Shanghai (China).

Keywords Urban agriculture · Creative people · Social entrepreneurship · Japan · China

1.1 Main Issues and Purpose

In general, when land prices rise with the development of cities, many agricultural lands are diverted and developed for non-agricultural uses, and agriculture is often regarded as unnecessary in cities. However, agriculture not only provides food production, but also provides a place for environmental conservation, landscape formation, agricultural experience, etc., and has a multifaceted function consisting of the triple bottom line aspects of economic efficiency, environmental conservation, and sociality (OECD 2001 etc.). And the relationship between these functions can be changed depending on the land tax system and zoning through city policy and city planning.

L. Kiminami et al., *Social Entrepreneurship, Social Business and the Multi-functionality of Urban Agriculture*, SpringerBriefs in Economics,
https://doi.org/10.1007/978-981-19-1762-2_1

Japanese cities are facing the problems of modern society such as aging, environment, land, housing and transportation. Against this background, the "Urban Agriculture Promotion Basic Law" was enacted in Japan in 2015, with the basic principles of demonstrating the multi-functionality of urban agriculture, good urban formation and coexistence of agriculture, and promotion of public understanding. On the other hand, in China, while the solution of the three agricultural problems and the construction of new rural areas are listed as important policy issues in the midst of rapid economic growth, the pressure of urban development is extremely strong.

However, recent studies have reported that residents' creativity is one of the important factors for urban development. For example, UNCTAD (2010) has set out an economic development strategy based on Creativity. In other words, it is possible to increase the amenities of the city (the attractive factor of agriculture from the aspect of enjoying multi-functionality) by promoting urban agriculture to attract the people with high creativity and productivity. And those highly creative people have the potential to solve social issues in the region and contribute to the sustainable development of the city by engaging in social business.

Creative class (CC) is a general term for people whose occupation is creating meaningful new forms (Florida 2002), and is considered as a source of growth. The relationship between creative class and the local economy has been shown to affect income levels, employment growth and entrepreneurship, although it depends on local conditions (eg., Hoyman and Faricy 2009; Gabe et al. 2007). It has also been reported that amenities such as the natural environment, tolerance and openness of the area promote the generation of creative classes (eg Boschma and Fritsch 2009; McGranahan and Wojan 2007; McGranahan et al. 2011). Regarding the creative class, research on the problems of concept/definition and indexing/measurement, generation factors and effects on the local economy has been mainly conducted, but there is almost no research related to agriculture. On the other hand, regarding agricultural amenities, there are studies on the evaluation of multifaceted functions such as Zasada (2011), Randall (2002), and Hiromasa and Fukazawa (1992), but there are few studies related to creative class.

Furthermore, studies on practical projects to solve urban social problems through urban agriculture (van Latesteijn and Rabbinge 2012 etc.) are mainly implemented in Europe, little full-scale academic research has been conducted on Asian countries including Japan. However, there are movements to meet the needs of urban residents for agricultural amenities in big cities of China, and new types of agricultural management bodies such as tourist farms, farm restaurants, and educational farms are emerging in some megacities. In order to realize the sustainable development goals (SDGs), research on social innovation for solving the problems such as food security, climate change and inequality/isolation etc. that megacities of the world are facing today is required. i.e. a process of changing people's mutual relationships and social mechanisms under new values with the aim of realizing a better society by solving social issues (Nonaka et al. 2014). However, little academic research has been conducted on social entrepreneurship in urban agriculture in Japan and China.

Therefore, the purpose of this book is to elucidate the mechanism that creative people with high level of social entrepreneurship solve regional issues through urban agriculture as a social business and contribute to sustainable urban development. It will mainly focus on the two megacities, Tokyo (Japan) and Shanghai (China).

1.2 Significance and Contributions

Research on urban agriculture has increased rapidly since the latter half of the 2010s, and more than 300 studies have been published annually in recent years (Fig. 1.1). In order to understand the characteristics of the research theme, we conducted an annual co-occurrence network analysis using the abstract information of each document from "Article" in Scopus. Based on the results, characteristic words such as 'farmer', 'activity', and 'water' were extracted in the latter half of the 2000s and the first half of the 2010s. Then, in the latter half of the 2010s, the characteristic words of 'community', and 'area' were extracted. Around 2020, 'food', 'production', 'development', and 'system' were extracted (Fig. 1.2). Therefore, in recent years (since the latter half of the 2010s), the accumulation of research with analytical viewpoints such as "community garden", "food system", and "city development" is progressing. Although important concepts of "social entrepreneurship" and "creative class" in this book have not emerged yet as trends in urban agriculture research, they may form important research areas in near future.

There are two main unique points in this book. The first is to build a new hypothetical model for explaining the mechanism of creating shared value (CSV) in urban agriculture for a sustainable urban development. The second is to analyze the mechanism of social entrepreneurship in urban agriculture from multi-layered (micro,

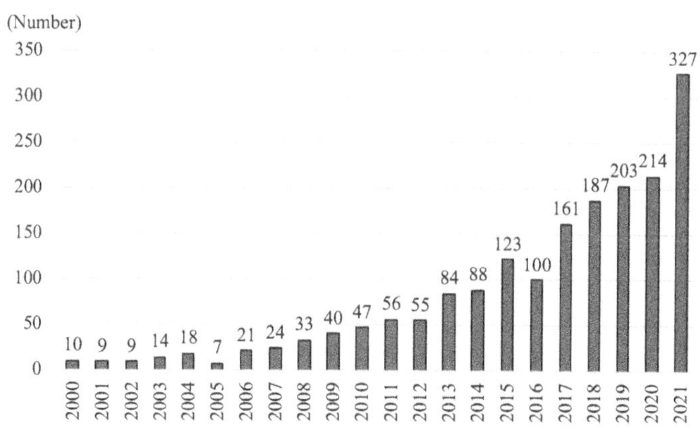

Fig. 1.1 Changes in the number of articles on "urban agriculture". *Source* Extracted from "Article" in Scopus. For the setting of search condition, "urban agriculture" was specified as the search term in the "Keywords" domain

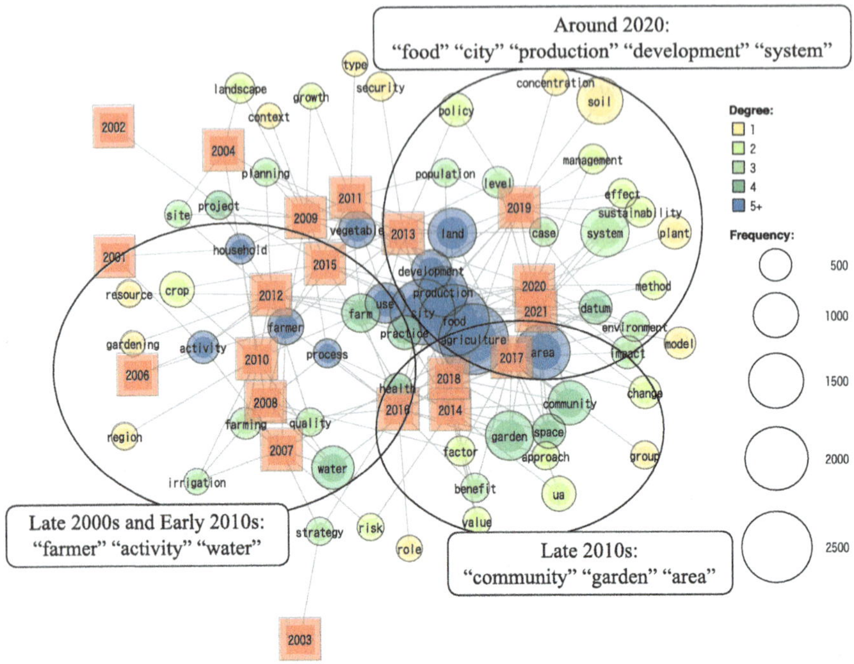

Fig. 1.2 Co-occurrence map of feature words by year of research topic. *Source* KH coder (Higuchi 2014) was used to make a co-occurrence network (top 200 co-occurrence relationships between words and publication years). Only the "noun" is specified as the analysis target. Due to the setting of the analysis software, all sentences of "Abstract" were converted to lowercase. In addition, words that do not form meaning in the context of this analysis (eg study, author, etc.) were deleted

meso, macro) and multifaceted (innovative and sustainable) aspects using a new analytical method (fusion of qualitative and quantitative analysis).

Research on social entrepreneurship has just begun to accumulate empirical research worldwide (Saebi et al. 2019 as a survey paper). A systematic approach to social entrepreneurship from the perspective of agricultural development is required (Christensen et al. 2019). This book is a new challenge to fuse the research on agriculture and rural development and that on urban sustainable development. It also makes a theoretical contribution to the existing entrepreneurship research (mainly based on the theory of business administration and economics) from the perspective of social business theory. Therefore, its academic significance is considered to be great.

Each chapter of this book is composed as shown in Fig. 1.3. In Chap. 2, we will summarize the achievements and issues of previous research by conducting a survey of the research on urban agriculture, social entrepreneurship, and creative class. Chapter 3 will describe a newly built theoretical framework and a mixed analytical method for hypothesis verification in this research. Chapter 4 will give an overview of the current state of urban agriculture in Japan and China and related policies and

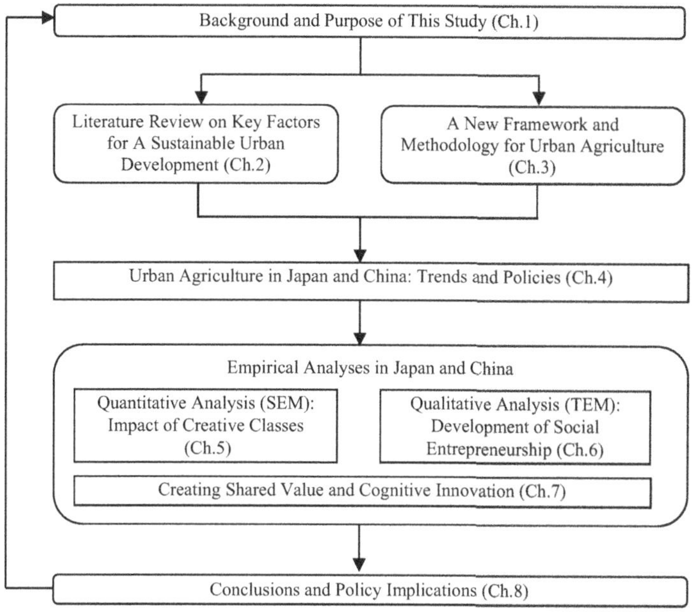

Fig. 1.3 Structure of this book

institutions. In Chap. 5, a quantitative analysis on the structure of awareness of urban residents toward multi-functionality of urban agriculture and social business will be undertaken by applying structural equation modeling (SEM) based on the results of a large-scale questionnaire survey targeting residents of Shanghai and Tokyo. In Chaps. 6 and 7, qualitative analyses of trajectory equifinality modeling (TEM) will be conducted to reveal the development mechanism of social entrepreneurship and creating shared value through cognitive innovation in urban agriculture both in China and Japan. Chapter 8 will present the conclusions and derive policy implications for practical efforts based on the analytical results in the previous chapters of this book. Issues left for further research will also be raised in the final chapter.

References

Boschma RA, Fritsch M (2009) Creative class and regional growth: empirical evidence from seven European countries. Econ Geogr 85(4):391–423

Christensen CM, Ojomo E, Dillon K (2019) The prosperity paradox: how innovation can lift nations out of poverty. Harper Business, New York

Florida R (2002) The rise of the creative class: and how it's transforming work, leisure, community and everyday life. Basic Books, New York

Gabe TM, Colby K, Bell KP (2007) Creative occupations, county-level earnings and the US rural-urban wage gap. Can J Reg Sci 30(3):393–410

Higuchi K (2014) Quantitative text analysis for social survey: toward succession and development of content analysis. Nakanishiya publishing, Kyoto (in Japanese)
Hiromasa Y, Fukazawa F (1992) Evaluations of farmland externality in the city area: hedonic price approach. Hokkaido J Agric Econ 2(1):27–35 (in Japanese)
Hoyman M, Faricy C (2009) It takes a village: a test of the creative class, social capital, and human capital theories. Urban Affairs Rev 44(3):311–333
McGranahan D, Wojan T (2007) Recasting the creative class to examine growth processes in rural and urban counties. Reg Stud 41(2):197–216
McGranahan DA, Wojan TR, Lambertz DM (2011) The rural growth trifecta: outdoor amenities, creative class and entrepreneurial context. J Econ Geogr 11:529–557
Nonaka I, Hirose A, Hirata T (2014) Social innovation. Chikura Shobo, Tokyo (in Japanese)
OECD (2001) Multifunctionality: towards an analytical framework. OECD Publishing, Paris
Randall A (2002) Valuing the outputs of multifunctional agriculture. Eur Rev Agric Econ 29(3):289–307
Saebi T, Foss NJ, Linder S (2019) Social entrepreneurship research: past achievements and future promises. J Manag 45(1):70–95
UNCTAD (2010) Creative economy report 2010. http://unctad.org/en/Docs/ditctab20103_en.pdf
van Latesteijn HC, Rabbinge R (2012) Wicked problems in sustainable agriculture and food security, the TransForum experience. Int Food Agribus Manage Rev 15(B):89–94
Zasada I (2011) Multifunctional peri-urban agriculture: a review of societal demands and the provision of goods and services by farming. Land Use Policy 28:639–648

Chapter 2
Literature Review

Abstract Literature review on some key factors for a sustainable urban development is conducted in this chapter. Creativity which is considered as the driving force of development has four aspects of 'Person', 'Process', 'Product', and 'Press' according to (Cropley et al (eds) Encyclopedia of creativity, Elsevier, 2011). In addition, Westlund et al. (Westlund et al, Handbook of research on entrepreneurship and creativity, Edward Elgar Publishing, Cheltenham, 2014) positioned creativity as an element of social capital (SC), and regarded it as the relationship between Press and Person. On the other hand, the rationale of multi-functionality of agriculture is explained in this chapter from the theory of Creating Shared Value (CSV), which is the idea that a company creates social value by working to solve social needs and social problems, and as a result, economic value is created (Porter and Kramer, Harv Bus Rev 89:62–77, 2011). It pointed out that no research has analyzed urban agriculture with multiple functions as a social business (SB) from the perspective of enhancing the creativity and sustainability of cities through the involvement of creative class so far.

Keywords Creativity · Creative class (CC) · Social capital (SC) · Multi-functionality of agriculture (MFA) · Social entrepreneurship (SE) · Social business (SB)

2.1 Creativity and Creative Class (CC)

Creativity is considered to be the driving force of development in various fields. The area related to creativity has traditionally been centered on the arts field, but it has been expanded to the economic field nowadays. For example, UNCTAD (2010) has set out an economic development strategy based on creativity.

According to Cropley (2011), there is a process behind a creative product to be created, and there is a person who executes that process. In addition, there are environmental pressures that promote or impede these activities. Therefore, creativity has four aspects: 'Person', 'Process', 'Product', and 'Press'. Although, Cerisola (2018) describes creativity using the terms of 'Content', 'Process', 'Outcome (Product)'

L. Kiminami et al., *Social Entrepreneurship, Social Business and the Multi-functionality of Urban Agriculture*, SpringerBriefs in Economics,
https://doi.org/10.1007/978-981-19-1762-2_2

and 'Source', and Westlund et al. (2014) use the terms of 'Personality', 'Process', 'Outcomes', and 'Milieus', they almost correspond to the above four aspects.

On the other hand, Florida (2002) defines creative class (CC) as "people in science and engineering, architecture and design, education, arts, music and entertainment whose economic function is to create new ideas, new technology, and new creative content." It was classified into super creative core, creative professionals, and artistically creative people. The 'super creative core' incorporates the likes of scientists, technicians, university professors, poets, novelists, artists, entertainers, actors, designers, and architects, while the 'creative professionals' are made up of professional occupations like tech or finance workers, lawyers, doctors, managers and so forth and tends to reside in areas with a high level of amenities. This approach has the effect of activating empirical research on 'creativity' because it is easy to identify CC, however, 'creativity' here was only defined from the aspect of Person (or Content).

In addition to the fact that there is no guarantee that Person (or Personality) and Process will match, occupation is only an inaccurate proxy variable for Personality. Therefore, not all people engaged in the so-called creative professions are creative (for example, creative thinking), and conversely, those engaged in so called non-creative professions are possible to be creative in terms of Process. Similar problems arise when defining CC in association with a creative industry. To address this issue, McGrahan and Wojan (2007) used the 'creative thinking' element of the 2004 O * NET content model to reconstruct occupations that are often creative in terms of Process as creative occupations. This can be seen as an attempt to match the 'Process' aspect of creativity with the 'Person' aspect and even with the profession. Kiminami et al. (2018, 2019) also conducted empirical analysis by treating people who emphasize creative thinking as CC in the same way.

Most of the empirical research on CC can be roughly divided into (1) research on factors that attract CC and (2) research on the impact of CC on the region. For studies on factors that attract CC, it has evolved on the basis of economic research on location selection and labor migration, as well as analysis of the effects of factors that attract CC such as Amenity, Tolerance, and Diversity (see Boschma and Fritsch 2009; Florida et al. 2008; Lovett and Beesley 2007; Rickman and Rickman 2011; Argent et al. 2013; Hatcher et al. 2011; McGranahan and Wojan 2007; McGranahan et al. 2011; Haisch and Klöpper 2015; Cerisola 2018; and Kiminami et al. 2018, 2019, 2020). In the process, there was a lively debate on whether employment opportunities or amenities were more important factors attracting CC. On the other hand, there was also a debate on causality (Østbye et al. 2018). In any case, these research results premise that the creativity of CC contributes to the economic development of the region, and became the policy basis for local governments to implement the conditions for guiding CC to the regions as a regional economic strategy.

Next, for studies on the impact of CC on the region, the focus of research in this area is on the role of CC in regional economic growth (see Gabe et al. 2007; Boschma and Fritsch 2009; Florida 2014; Mellander et al.2014; Westlund and Calidoni 2013; Deller et al. 2001; Argent et al. 2013). Most of them econometrically analyzed the relationship between CC identified by occupational classification and variables of economic outcomes such as income and employment, and clarified the contribution of CC to regional economic growth. In a series of studies (see Batabyal 2021; Batabyal et al. 2019; Batabyal and Yoo 2020), CC is used as the Creative Capital and theoretically analyzed using a regional economic growth model based on the framework of microeconomics. It also formulated the optimal number of CC and the optimal level of financial support for CC. In addition, CC is compared to human capital (HC) and there is a debate over which is the main driver of regional economic growth. Using structural equation models and path analysis, Florida et al. (2008) showed that human capital and CC influence regional development in different ways.

On the other hand, many studies question the importance of CC as a growth factor for the regional economy (Faggian et al. 2017). Erro-Garcés (2020) stated that social value is created as creativity, innovation and entrepreneurship develop together, based on the results of three case studies of leading social entrepreneurs in Spain. In other words, the value created by creative innovation develops as a new business for entrepreneurs and becomes a new activity that brings value to many people.

It should be considered that the CC originally proposed by Florida (2002) are the people who comprehensively satisfied various aspects of creativity and brought about a great change in society. In that sense, many CC studies so far had only analyzed the relationship between limited aspects of creativity, and the analysis itself is often synonymous and tautological, and the analysis target has been biased. Therefore, future research on CC will require a comprehensive capture of creativity related to the sustainable development of the society from Person, Process, Product, and Press.

2.2 Whether Social Capital (SC) Matters for the CC?

Social capital has multiple definitions and one of the representative definitions is "features of social organization, such as trust, norms, and networks that can improve the efficiency of society by facilitating coordinated actions" (Putnam 1993), which is generally categorized into bonding social capital and bridging social capital by the differences in function of connecting actors. However, bonding social capital is a connection among homogeneous actors within social organizations, which internally generate trust, cooperation and unity. On the other hand, bridging social capital is a network among heterogeneous actors and social organizations, which has the functions of sharing and exchanging information and building consensus among the groups with diverse interests.

Which theory has stronger explanatory power among social capital (SC), human capital (HC) and creative class (CC) on regional economic development has been discussed for some years (Florida 2003; Hoyman and Faricy 2009), despite the debates on the question of how to define the core of these factors. Especially, there are arguments both for and against the role of Bonding SC in the regional economic development. Although Westlund and Calidoni (2013) could not find a clear impact of civil society on regional development from the empirical study on the relationship between the CCs and social capital in Japan, Rich (2012) pointed out that small cities attempting to achieve economic growth and population stability should focus on their strengths: city livability and the thick social ties that maintain communities. Westlund et al. (2014) also gave a similar view on the relationship between Creativity, Entrepreneurship, and Innovation in the context of economic development, by positioning creativity as an element of social capital, and regarded it as the relationship between Press and Person. Furthermore, based on the questionnaire survey targeting at the residents in Tokyo, Kiminami et al. (2018) clarified that the CCs who are "thinking creatively" also have a high level of social capital and are favorable to diversified lifestyles that are thought to contribute to the development of cities.

On the other hand, using the data of 16,253 participants from 29 provinces /municipalities in China based on China Labor-force Dynamic Survey (CLDS), Li (2017) showed that the positive role of generalized trust in promoting success was less sensitive to the influence of societal residential mobility. In this research, we will testify the relationships among multi-functionality of urban agriculture (MFA), the CCs and social capital (SC) in Shanghai and Tokyo to make contribution to the debates on the issue.

2.3 Social Entrepreneurship, Social Innovation and Social Business

Social entrepreneurship (SE) is defined by its two constituent elements: a prime strategic focus on social impact and an innovative approach to achieving its mission (Nicholls 2006). Social entrepreneurship (SE) activity is facilitated by dedicated finance programs, which include community investment, program-related investment, and dedicated legal status. There is no "one-size-fits-all" blueprint for social entrepreneurship because of the vast differences in social venturing prevalence as well as legal and regulatory frameworks, access to financial resources, markets, and training (Terjesen et al. 2016). However, Stephan et al. (2015) highlighted four factors that affect SE: institutional voids, institutional support, post-materialism cultural values and socially supportive culture norms (weak-tie social capital) and pointed out that the promotion of SE is greatly influenced by government support and values based on the results of quantitative analysis using Global Entrepreneurship Monitor (GEM) data. In addition, based on the review results of 395 previous studies on SE, Saebi et al. (2019) pointed out that an approach to understand social entrepreneurship from the three levels of individuals, organizations, and institutions and to analyze their manifestation and change processes is necessary.

On the other hand, Montgomery et al. (2012) regards social entrepreneurship as belonging to a group or organization rather than to an individual. Specifically, collective social entrepreneurship is the solution to social problems using a business approach through the cooperation of various actors (collective action), which is considered to be the source of collective impact. In addition, Mato-Santiso and

Rey-Garcia 2019) proposes an analysis framework for collective social entrepreneurship and conducts case analysis while partially expanding the framework for collective impact. In particular, the manifestation of collective social entrepreneurship is characterized by focusing on the fact that it exerts collective impact and causes system changes as the construction of an ecosystem of shared value. Kovanen (2021) reviewed previous studies focusing on aspects of collaboration in social entrepreneurship and pointed out that there are many studies on vague and non-critical although they relied on three theories of community and government, resource and employment, and complex network collaboration. In addition, Pel et al. (2020) conducted research focusing on how the process of social innovation causes transformative change. As a result of analyzing 20 types of network organizations related to social innovation, 12 propositions have been derived. Furthermore, Strasser et al. (2020) found that three institutional aspects of structural and cultural embeddedness, widespread and coherent influence, and persistence and evolving reproduction are important in creating a transformative impact and promote the emergence of social innovation.

Social innovation is defined as the process of innovation that solves social problems at a gradual, institutional and destructive level for sustainable regional development (e.g., dealing with social needs, market and government failure) (Nicholls and Murdock 2012). Research on social innovation has two streams, 'Practical Stream' and 'Critical Stream' (Moulaert and MacCallum 2019, p. 36). The former is an approach to study entrepreneurship and management issues for solving the social problems in real-life, and the latter is an approach to study system and institutional issues for group empowerment, and it is necessary to integrate these in a complementary manner.

Although the accumulation of empirical and quantitative analysis on SE and social business (SB: the entities that engage in profitable businesses using business methods for the purpose of solving social issues) has rapidly advanced since the beginning of the 2010s, few studies have evaluated and analyzed urban agriculture from the viewpoint of them (for example, Dimitri et al. 2016; Specht et al. 2016, etc.). Furthermore, there is no research that analyzes urban agriculture with multiple functions as a social business and analyzes it from the perspective of enhancing the creativity and sustainability of cities through the involvement of creative classes.

2.4 Multi-functionality of Urban Agriculture

Multi-functionality and multiple functions of agriculture are terms used to indicate generally that agriculture can produce various non-commodity outputs in addition to food (OECD 2000). Multi-functionality of agriculture is composed of economic, environmental, and social aspects that supplies not only food but also services such as environmental conservation, good scenery and farm experience (see Table 2.1). Zasada (2011) has offered a survey of effects in each area, though it has been multiple, there has been no agreed-upon definition of what they are, but these varied effects

Table 2.1 Multi-functionality of urban agriculture

Function	Contents
Production	Basic functions such as production of foods
Communication	Enjoyment of culture and creation of communication through exchange between citizens and between citizens and farmers
Welfare	Prevention of aging through agricultural work, healing effects of plants, gardening therapy
Education	Emotional and environmental education through nature and agriculture and learning agriculture and forestry
Recycling	Organically grown vegetables by turning kitchen garbage into organic fertilizer
Environmental protection	Preservation of biological resources and natural environment
Creation of landscapes	Creation of pleasant landscapes, scenery of Japanese fields and scenery through which people can enjoy changes of the seasons
Disaster prevention	Disaster evacuation sites and routs, green spaces for disaster prevention, spread of fire, spaces for temporary housing
History and culture	Preservation of groves of village shrines and continuation of harvest festivals
Support for building residential lands	Promotion of building residential lands, provision of gardens and vegetable gardens that support good rural living
Withholding or controlling of urbanization	Temporary withholding or controlling urbanization for a certain period of time

range broadly across society and the environment. By comparing a number of valuation methods, Randall (2002) pointed out the difficulties of appropriate evaluation of these multiple functions.

In Japan, many studies have adopted the Hedonic approach as applicable to urban agriculture, such as Hiromasa and Fukazawa (1992), Maruyama et al. (1995), Terawaki (1997) and Hara and Kato (2005). However, these studies showed that the existence of agriculture in cities drove down the prices of surrounding land. As for the research on the evaluation of amenities in cities, Shimizu et al. (2014) analyzed the correlation between urban amenities and rents as the price of housing using the same Hedonic approach.

Kiminami and Kiminami (2006, 2007) analyzed the awareness of urban residents regarding urban agriculture in Tokyo and Shanghai. The results showed that residents' subjective impressions towards urban agriculture influence their consciousness of settlement in large cities. The study evaluated the multi-functionality of agriculture by showing that urban agriculture enhances the quality of urban life and the settlement of its residents. Furthermore, Peng et al. (2015) evaluated the multiple functions of urban agriculture in Beijing using proxy data for the economic, social and environmental effects of agriculture.

As for the roles of urban agriculture, Artmann and Sartison (2018) systematically reviewed research papers on nature-based solution of urban agriculture since 2010 (166 editions mainly in Global North), and clarified that urban agriculture contributes to the solution of ten social issues that cities are facing. It also pointed out that when introducing urban agriculture, both drivers and inhibitors must be considered across social, institutional, economic, technical, geographical and ecological aspects. However, there is no research that captures creative class (CC) as an entity that promotes urban agriculture and overcomes the obstacles.

As for the relationship between urban agriculture and the creative class, Woldoff et al. (2011) clarified that urban residential preferences of college students in creative majors are Bohemian including Organic Food Market. Furthermore, results from the multi-functionality of urban farming in Vancouver also clarified that urban agriculture plays a role in developing inclusive, resilient, safe and diverse food systems (Valley and Wittman 2019). However, researches on the relationship between regional factors attracting the CCs focusing on the people who are thinking creatively (TC) and multi-functionality of urban agriculture are insufficient.

On the other hand, the Cabinet Office of Japan (2016) defined social impact as "social and environmental outcomes resulting from the business or activity, including short-term and long-term changes." It also proposed to quantitatively and qualitatively grasp the outcomes, judge the value of the business and activities, and make improvements as necessary. Furthermore, in assessing social impact, it pointed out the importance of achieving the two objectives of (1) fulfilling accountability and (2) learning and improving (a wide range of stakeholders' participation, collaboration and evaluation). In addition, Stroh (2015) argued that system thinking is important for achieving collective impact, and in particular, by sustaining the key collaborative efforts for change over a long period of time, a big difference will be made in the entire system.

2.5 Creating Shared Value (CSV) in Urban Agriculture

Although many studies have pointed out the existence of multi-functionality of agriculture, it is necessary to generate economic value and social value at the same time. The rationale is the theory of Creating Shared Value (CSV), which is the idea that a company creates social value by working to solve social needs and social problems, and as a result, economic value is created (Porter and Kramer 2011). Unlike the conventional way of thinking about CSR and social contribution activities, CSV is positioned at the center rather than around corporate activities, and it is significant that it changes the way of thinking about business itself. Kramer and Pfitzer (2016) pointed out that many companies adopted the CSV strategy but are not successful because of working through their own business activities. However, Kamer and Pfitzer (2016) pointed out that it is not easy for companies to make CSV successful in the face of uncontrollable barriers such as changes in demand, cultural norms and government regulations. They added that it is necessary to involve governments, NGOs and communities to form a shared value ecosystem and create collective impact. In particular, it is important to exert a collective impact through collaboration between social enterprises and various stakeholders (Kania and Kramer 2011). In addition, de los Reyes et al. (2017) stated that in order for CSV to work, it is necessary to establish an Ethical management framework that complements it. Furthermore, Menghwar and Daood (2021) classified the factors affecting the realization of CSV into external and internal ones.

In terms of external factors, the first one is considered to be the system. Liang and Renneboog (2017) found that there is a high correlation between the legal origin behind the economic activities of each country and CSR performance according to an international comparative study. In other words, differences in systems are considered to affect the performance in CSV. However, systems here are not only considered as the laws and the regulations by the state, but also the rules that people are following normatively and cognitively (whether formal or informal), and social capital is also included. And the second one is the trends of competitors (Competitors' approach). Cao et al. (2019) empirically showed that competitors' CSR efforts affect their decision-making regarding the introduction of CSR. Similarly, the implementation of CSV by a company may depend on the strategies of its competitors. Therefore, the CSV strategy in urban farming is influenced by the reality of CSV of other farming and non-farming companies competing for resources, products and services. The third external factor is customer's behavior. Ham et al. (2020) found that an eating out company's CSV strategy affects the effect of consumers on raising brand attitudes through awareness of the company's social contribution, resulting in increased demand for the company. Therefore, the recognition and evaluation of the economic and multi-functionality of urban agriculture from consumers, as well as from local stakeholders of agricultural products and services will determine the CSV strategy of urban agriculture.

In terms of internal factors, the first one is considered to be the emergent strategy of a company. Through the case analysis of Wal-Mart, Spicer and Hyatt (2017) clarified that the emergent strategy plays an important role in creating shared value. In other words, entrepreneurship, which creates an emergent strategy, is considered to contribute significantly to the success of CSV. The second one is the visionary leadership. Vaidyanathan and Scott (2012) pointed out that a vision of leadership and a clear strategic decision-making are important for implementing CSV, and the motivation for enterprises to adopt CSV is related to the morality in leadership (Chen et al. 2018), awareness of survival and strong ethics (Kim et al. 2019). In order to realize CSV in urban agriculture, it is important for agricultural managers to be aware of the management philosophy and the current state of the society. The third one is cognitive capability. Barr et al. (1992) revealed that changes in mental models of manager (cognitive maps in empirical analysis) promote strategic innovation in organizations. Corner and Pavlovich (2016) regarded metacognition as essential for individuals to overcome habitual decision-making patterns and improve organizational value creation processes. Metacognition can encourage innovation especially when conflicts need to be overcome (Lorenz et al. 2018), such as in the simultaneous creation of social/environmental and economic value in hybrid organizations (Yaari et al. 2021). Furthermore, as a precondition for strategic innovation of a company, Rajagopalan and Spreitzer (1997) and Kida (2007) focused on the changes in organizational knowledge structure including manager's cognitive structure from a cognitive perspective.

Therefore, in this research, we consider creating shared value (CSV) as the rationale for urban agriculture to bring out multiple functions while conducting economic business (creating economic value and social value at the same time). Furthermore, cognitive innovation of social entrepreneurship in urban agriculture will lead to the success of CSV by involving local stakeholders.

References

Argent N, Tonts M, Jones R, Holmes J (2013) A creativity-led rural renaissance? Amenity-led migration, the creative turn and the uneven development of rural Australia. Appl Geogr 44:88–98

Artmann M, Sartison K (2018) The role of urban agriculture as a nature-based solution: a review for developing a systemic assessment framework. Sustainability 10(6):1937

Barr PS, Stimpert JL, Huff AS (1992) Cognitive change, strategic action and organizational renewal. Strateg Manag J 13:15–36

Batabyal AA (2021) Monopoly versus individual welfare when a local public good is used to attract the creative class. Int Reg Sci Rev 44(5):605–614

Batabyal AA, Kourtit K, Nijkamp P (2019) Using local public goods to attract and retain the creative class: a tale of two cities. Reg Sci Policy Pract 11:571–581

Batabyal AA, Yoo SJ (2020) Using a local public good to attract representative creative class members: the inefficient equilibrium case. Theoret Econ Lett 10:40–46

Boschma RA, Fritsch M (2009) Creative class and regional growth: empirical evidence from seven European countries. Econ Geogr 85(4):391–423

Cabinet Office of Japan (2016) Toolkit for social impact assessment (Investigation on the development for human resource and organizational capacity through the practice of social impact assessment). https://www.npo-homepage.go.jp/uploads/h28-social-impact-chousa-report-03.pdf (in Japanese)

Cao J, Liang H, Zhan X (2019) Peer effects of corporate social responsibility. Manage Sci 65:5487–5503

Cerisola S (2018) Creativity and local economic development: the role of synergy among different talents. Pap Reg Sci 97(2):199–215

Chen YRR, Hung-Baesecke CJF, Bowen SA, Zerfass A, Stacks DW, Boyd B (2018) The role of leadership in shared value creation from the public's perspective: a multi-continental study. Publ Relations Rev 46:1–11

Corner PD, Pavlovich K (2016) Shared value through inner knowledge creation. J Bus Ethics 135:543–555

Cropley AJ (2011) Definitions of creativity. In: Runco MA, Pritzker SR (eds) Encyclopedia of creativity. Elsevier, pp 511–524

Deller SC, Tsai T (S), Marcouiller DW, English DBK (2001) The role of amenities and quality of life in rural economic growth. Am J Agric Econ 83(2):352–365

de los Reyes G, Scholz M, Smith NC (2017) Beyond the "Win-Win": creating shared value requires ethical frameworks. California Manage Rev 59(2):142–167

Dimitri C, Oberholtzer L, Pressman A (2016) Urban agriculture: connecting producers with consumers. Brit Food J 118(3):603–617
Erro-Garcés A (2020) Creativity and emotions as drivers for social entrepreneurship. J Social Entrepreneurship 11(3):300–316
Faggian A, Partridge M, Malecki EJ (2017) Creating an environment for economic growth: creativity, entrepreneurship or human capital? Int J Urban Reg Res 41(6):997–1009
Florida R (2002) The rise of the creative class: and how it's transforming work, leisure, community and everyday life. Basic Books, New York
Florida R (2003) Cities and creative class. City Community 2(1):3–19
Florida R (2014) The creative class and economic development. Econ Dev Q 28(3):196–205
Florida R, Mellander C, Stolarick K (2008) Inside the black box of regional development: human capital, the creative class and tolerance. J Econ Geogr 8:615–649
Gabe TM, Colby K, Bell KP (2007) The effects of workforce creativity on earnings in U.S. counties. Agric Res Econ Rev 36(1):71–83
Haisch T, Klöpper C (2015) Location choices of the creative class: does tolerance make a difference? J Urban Aff 37(3):233–254
Ham S, Lee S, Yoon H, Kim C (2020) Linking creating shared value to customer behaviors in the food service context. J Hosp Tour Manag 43:199–208
Hara N, Kato K (2005) Hedonic evaluation of externalities of agricultural lands in big cities. Kagawa Univ Econ Rev 78(1):33–49 (in Japanese)
Hatcher W, Oyerb M, Gallardoc R (2011) The creative class and economic development as practiced in the rural U.S. south: an exploratory survey of economic development professionals. Rev Region Stud 41(2, 3):139–159
Hiromasa Y, Fukazawa F (1992) Evaluations of farmland externality in the city area: Hedonic price approach. Hokkaido J Agric Econ 2(1):27–35 (in Japanese)
Hoyman M, Faricy C (2009) It takes a village: a test of the creative class, social capital, and human capital theories. Urban Affairs Rev 44(3):311–333
Kania J, Kramer M (2011) Collective impact. Stanford Social Innov Rev Winter 2011. https://ssir.org/articles/entry/collective_impact
Kida M (2007) Cognitive innovation: visualization of cognitive change and knowledge and introduction of text mining to organizational science. Hakuto-Shobo Publishing, Tokyo (in Japanese)
Kim RC, Saito A, Avvari VM (2019) Interpretation and integration of 'creating shared value' in Asia: implications for strategy research and practice. Asian Bus Manag 19:379–406
Kiminami L, Kiminami A (2006) Sustainability of agriculture and urban quality of life in Japan: economic efficiency, sociality and environment protection. Stud Region Sci 36(2):305–321
Kiminami L, Kiminami A (2007) Sustainability of urban agriculture: a comparative analysis of Tokyo and Shanghai. Stud Region Sci 37(2):585–597
Kiminami L, Kiminami A, Furuzawa S (2018) Impacts of multi-functionality of urban agriculture on the CCs in Japan. Asia-Pacific J Region Sci 2(2):507–527
Kiminami L, Furuzawa S, Kiminami A (2019) Impacts of multi-functionality of urban agriculture on the creative classes in global mega city: focusing on Shanghai in China. Asia-Pacific J Region Sci 3(2):487–515
Kiminami L, Furuzawa S, Kiminami A (2020) Social entrepreneurship and social business associated with multiple functions of urban agriculture in Japan. Asia-Pacific J Region Sci 4(2):521–552
Kovanen S (2021) Social entrepreneurship as a collaborative practice: literature review and research agenda. J Entrepreneurship Manage Innov 17(1):97–128
Kramer MR, Pfitzer MC (2016) The ecosystem of shared value. Harv Bus Rev 94(10):80–89
Liang H, Renneboog L (2017) On the foundations of corporate social responsibility. J Financ 72:853–910
Li LMW (2017) Social class, social capital and residential mobility in China. Soc Indic Res 132(2):1117–1129

Lorenz MP, Ramsey JR, Richey RG (2018) Expatriates' international opportunity recognition and innovativeness: the role of metacognitive and cognitive cultural intelligence. J World Bus 53(2):222–236
Lovett C, Beesley K (2007) Where to live? The residential preferences of Canada's creative class. Prairie Persp 10:47–55
Maruyama A, Sugimoto Y, Kikuchi M (1995) Farmland and green space in a residential area: hedonic evaluation of urban amenities in Chiba City. J Rural Econ 67(1):1–9 (in Japanese)
Mato-Santiso V, Rey-Garcia M (2019) Collective social entrepreneurship for sustainable impact: a framework proposal. World Rev Entrepreneurship Manage Sustain Devel 15(1/2):10–28
McGranahan D, Wojan T (2007) Recasting the creative class to examine growth processes in rural and urban counties. Reg Stud 41(2):197–216
McGranahan DA, Wojan TR, Lambertz DM (2011) The rural growth trifecta: outdoor amenities, creative class and entrepreneurial context. J Econ Geogr 11:529–557
Mellander C, Florida R, Asheim BT, Gertler M (eds) (2014) The creative class goes global. Routledge, Abington
Menghwar PS, Daood A (2021) Creating shared value: a systematic review, synthesis and integrative perspective. Int J Manag Rev 23(4):466–485
Montgomery AW, Dacin PA, Dacin MT (2012) Collective social entrepreneurship: collaboratively shaping social good. J Bus Ethics 111:375–388
Moulaert M, MacCallum D (2019) Advanced introduction of social innovation. Edward Elgar Publishing, Cheltenham
Nicholls A (ed) (2006) Social entrepreneurship: new models of sustainable social change. Oxford University Press, Oxford
Nicholls A, Murdock A (2012) The nature of social innovation. In: Nicholls A, Murdock A (eds) Social innovation: blurring boundaries to reconfigure markets. Palgrave Macmillan, London, pp 1–30
OECD (2000) Valuing rural amenities. OECD Publishing, Paris
Østbye S, Moilanen M, Tervo H, Westerlund O (2018) The creative class: do jobs follow people or do people follow jobs? Reg Stud 52(6):745–755
Pel B, Haxeltine A, Avelino F, Dumitru A, Kemp R, Bauler T, Kunze I, Dorland J, Wittmayer J, Jørgensen MS (2020) Towards a theory of transformative social innovation: a relational framework and 12 propositions. Res Policy 49(8):1–13
Peng J, Liu Z, Liu Y, Hu X, Wang A (2015) Multifunctionality assessment of urban agriculture in Beijing City. China Sci Total Environ 537:343–351
Porter ME, Kramer MR (2011) The big idea: creating shared value (How to reinvent capitalism—and unleash a wave of innovation and growth). Harv Bus Rev 89(1–2):62–77
Putnam R (1993) Making democracy work: civic traditions in modern Italy. Princeton University Press, Princeton
Rajagopalan N, Spreitzer GM (1997) Toward theory of strategic change: a multi-lens perspective and integrative framework. Acad Manag Rev 22(1):48–79
Randall A (2002) Valuing the outputs of multifunctional agriculture. Eur Rev Agric Econ 29(3):289–307
Rich MA (2012) From coal to cool: the creative class: social capital, and the revitalization of Scranton. J Urban Aff 35(3):365–384
Rickman DS, Rickman SD (2011) Population growth in high-amenity nonmetropolitan areas: what's the prognosis? J Reg Sci 51(5):863–879
Saebi T, Foss NJ, Linder S (2019) Social entrepreneurship research: past achievements and future promises. J Manag 45(1):70–95
Shimizu C, Yasumoto S, Asami Y, Clark TN (2014) Do urban amenities drive housing rent? Grant-in-aid for scientific research (S) real estate markets, financial crisis, and economic growth: an integrated economic approach working paper series No 9
Specht K, Weit T, Swoboda K, Siebert R (2016) Socially acceptance urban agriculture businesses. Agron Sustain Dev 36(17):1–14

Spicer A, Hyatt D (2017) Walmart's emergent low-cost sustainable product strategy. Calif Manage Rev 59:116–141
Stephan U, Uhlaner LM, Stride C (2015) Institutions and social entrepreneurship: the role of institutional voids, institutional support, and institutional configurations. J Int Bus Stud 46:308–331
Strasser T, de Kraker J, Kemp R (2020) Three dimensions of transformative impact and capacity: a conceptual framework applied in social innovation practice. Sustainability 12:4742
Stroh DV (2015) Systems thinking for social change. Chelsea Green Publishing, White River Junction
Terawaki T (1997) Measuring positive and negative externalities of urban farmland. J Rural Plan Assoc 13(3):216–227 (in Japanese)
Terjensen S, Bosma N, Stam E (2016) Advancing public policy for high-growth, female and social entrepreneurs. Public Adm Rev 76(2):230–239
UNCTAD (2010) Creative economy report 2010
Vaidyanathan L, Scott M (2012) Creating shared value in India: the future for inclusive growth. J Decis Makers 37:108–113
Valley W, Wittman H (2019) Beyond feeding the city: the multifunctionality of urban farming in Vancouver, BC. City Cult Soc 16:36–44
Westlund H, Calidoni F (2013) Regional development and the creative class in Japan. In: Mellander C Florida R, Asheim BT, Gertler M (eds) The creative class goes global. Routledge, Abington, pp 262–282
Westlund H, Andersson M, Karlsson C (2014) Creativity as an integral element of social capital and its role in economic performance. In: Sternberg R, Krauss G (eds) Handbook of research on entrepreneurship and creativity. Edward Elgar Publishing, Cheltenham, pp 60–96
Woldoff RA, Decola T, Litchfield RC (2011) The aspirational creative class: urban residential preferences of college students in creative majors. City Cult Soc 2:75–83
Yaari M, Blit-Cohen E, Savaya R (2021) Hybrid organizational culture: the case of social enterprises. Journal of Social Entrepreneurship 12(2):291–313
Zasada I (2011) Multifunctional peri-urban agriculture: a review of societal demands and the provision of goods and services by farming. Land Use Policy 28:639–648

Chapter 3
A New Framework and Methodology for Urban Agriculture

Abstract In this chapter, we construct a hypothetical model of the mechanism of creating shared value (CSV) in urban agriculture based on the review of previous studies in Chap. 2 and the theoretical examination. A mixed methods approach to combine a quantitative analysis of structural equation modeling (SEM) and a qualitative analysis of trajectory equifinality modeling (TEM) is designed for hypothesis verification in this research. The advantage of the mixed methods approach is expected to elucidate the complex problems that cannot be clarified by either quantitative or qualitative methods alone. That is, the multi-functionality of urban agriculture and the social institutions influence the cognition and behavior of stakeholders in urban agriculture, and innovation in collective cognition through organizational learning among stakeholders will enhance social entrepreneurship and improve the sustainability of the city. And these effects will be feedbacked to the multi-functionality of urban agriculture, the social institutions, and the cognition and behavior of stakeholders.

Keywords Hypothetical model · Creating shared value (CSV) · Mixed methods approach · Structural equation modeling (SEM) · Trajectory equifinality modeling (TEM)

3.1 Theoretical Examination

In accordance with Kiminami and Kiminami (2006) and Kiminami et al. (2018), the multi-functionality of urban agriculture is regarded as supply of private goods and public goods with externality. And, the utility function of the residents is expressed as follows:

$$U = U(C, N, A, G, MFA; SB) \tag{3.1}$$

Here, U is the utility function of residents; C is the consumer goods; N is the housing service; A is the pure amenity; G is the public goods; and MFA is the

L. Kiminami et al., *Social Entrepreneurship, Social Business and the Multi-functionality of Urban Agriculture*, SpringerBriefs in Economics,
https://doi.org/10.1007/978-981-19-1762-2_3

multi-functionality of Urban Agriculture; and SB is the social business. If the multi-functionality of urban agriculture is classified into three categories of economic function (EC), environmental function (EN) and social function (SO), Eq. (3.2) is expressed as follows:

$$U = U\{C, N, A, G, MFA(EC, EN, SO); SB\} \tag{3.2}$$

When the utility functions of the CC (creative class) and non-CC are UC and UN, respectively, and the multi-functionality of urban agriculture has a positive influence on the utility of residents, we obtain the following conditional expression:

$$\partial UC/\partial MFA > 0 \tag{3.3a}$$

$$\partial UN/\partial MFA > 0 \tag{3.3b}$$

Furthermore, if CC strongly prefers the multi-functionality of agriculture to non-CC, the following formula holds.

$$\partial UC/\partial MFA > \partial UN/\partial MFA \tag{3.4}$$

Also, if CC prefers social capital more strongly than non-CC, the following formula holds.

$$\partial UC/\partial SC > \partial UN/\partial SC \tag{3.5}$$

In addition, promoting Social Business (SB) will eliminate failures in the market and government sector and effectively enhance the utility of urban residents through market and community mechanisms. The utility of the inhabitants is enhanced.

$$\partial UC/\partial SB > 0 \tag{3.6}$$

Here, the supply level of social business is expressed as a function of social capital and creative class as follows.

$$SB = SB\,(SC, CC) \tag{3.7}$$

Social business is a new category of business that aims to solve social problems through business. Therefore, if the stronger the relationship of trust and network between the residents and the stronger the creative thinking, the higher the level of social business. They are expressed as follows.

$$\partial SB/\partial SC > 0 \tag{3.8a}$$

$$\partial SB/\partial CC > 0 \tag{3.8b}$$

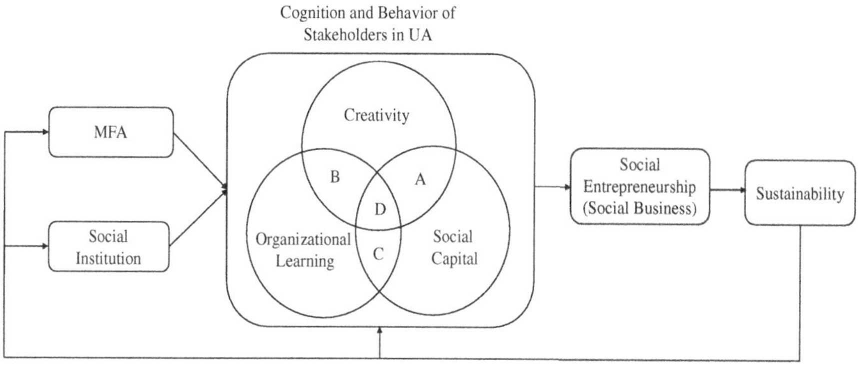

Fig. 3.1 A new conceptual framework: creating shared value in urban agriculture

Based on the review of previous studies in Chap. 2 and the above theoretical examination, a hypothetical model of the mechanism of creating shared value (CSV) in urban agriculture is constructed and shown in Fig. 3.1. That is, the multi-functionality of urban agriculture and the social institutions influence the cognition and behavior of stakeholders in urban agriculture, and innovation in collective cognition through organizational learning among stakeholders will enhance social entrepreneurship and improve the sustainability of the city. And these effects will be feedbacked to the multi-functionality of urban agriculture, the social institutions, and the cognition and behavior of stakeholders.

Here, we suppose that the cognitive innovation of stakeholders in urban agriculture will be most promoted when the three conditions of creativity, social capital, and organizational learning are met (area D). In other words, the higher the level of each three elements, the more social entrepreneurship develops through the radical cognitive innovation of stakeholders. In addition, in the cases when any of the social capital and organizational learning (area A), creativity and organizational learning (area B), and social capital and organizational learning (area C) is high, cognitive innovation is gradual, and level of social entrepreneurship is moderate. On the other hand, if the level of three factors is low, or if the level of only one factor is high, a dilemma of cognitive innovation will arise.

In Chap. 5 and subsequent chapters of this book, we will clarify the mechanism of creating shared value in urban agriculture shown in Fig. 3.1 by using a mixed methods approach. Mixed methods approach is broadly defined as the research that investigators collect and analyze data, integrate findings, and draw inferences using both qualitative and quantitative methods in a single study or program of inquiry (Tashakkori and Creswell 2007). The advantage of the mixed methods approach is that by appropriately combining quantitative and qualitative methods, it is expected to elucidate complex problems that cannot be clarified by either quantitative or qualitative methods alone. In this research, we will combine structural equation modeling (SEM) as a quantitative method and trajectory equifinality modeling (TEM) as a qualitative method to clarify the process of creating shared value in urban agriculture.

3.2 Explanation of the Methods

3.2.1 Mixed Methods (MM)

In general, the main purpose of research in the social sciences is to elucidate the phenomena that occur in the relationship between humans and society, but more practical research is required when aiming to solve social problems. In conventional research, either the "qualitative method" in which phenomena are understood by describing them in detail and with precision, or the "quantitative method" in which phenomena are viewed in terms of quantity and understood through statistical analysis, has been adopted. And the conceptualization and theorization of phenomena have been promoted by each research method. However, to elucidate complex phenomena and issues, the data to be collected and the methods of analysis might be more complex, but it is not easy to collect and analyze data with different characteristics. For this reason, mixed methods are gaining attention today as a research method that integrates qualitative and quantitative research. In the latter half of the 1980s, discussions began on the construction of mixed methods as a system that integrates qualitative and quantitative research, and since the 1990s, research has become active mainly in the fields of medicine, nursing, education, and welfare. Mixed methods have also been applied to issues related to social and economic development, starting with Bamberger (2000).

There are a variety of ways to integrate quantitative and qualitative research, and many typologies have been proposed for mixed methods research designs. Creswell and Plano Clark (2017) organized their mixed methods research designs into three core models: the convergent design, the explanatory sequential design, and the exploratory sequential design. In the convergent design, the collection and analysis of quantitative and qualitative data are conducted independently, and the results of the analysis are combined or compared. Therefore, a relationship in which qualitative research enhances or supplements quantitative research is expected. In the explanatory sequential design, quantitative data are collected and analyzed first, and then qualitative data are collected and analyzed to explain the results. Thus, it is expected that the qualitative data will explain the results of the quantitative research, or that the results of the quantitative research will be used to select the subjects of the qualitative research. In the exploratory sequential design, issues are first explored through qualitative data collection and analysis, and then quantitative data collection and analysis for the development of measurement scales and interventions are conducted based on the results. Therefore, quantitative data is expected to validate the generalization of findings obtained in qualitative research.

Our design for mixed methods approach in this research is basically considered as a convergent design in the sense that quantitative and qualitative researches are applied independently to the process of CSV in urban agriculture as shown in Fig. 4.1. However, it has an aspect of explanatory sequential design in which variables used in the quantitative research are based on the results of qualitative research conducted in the past. It also includes exploratory sequential design in which qualitative research analyzes the historical changes in social entrepreneurs and the feedback of CSV processes based on the analytical results of quantitative research.

3.2.2 Structural Equation Modeling (SEM)

Structural equation modeling (SEM) is an analysis in which the relationships among a large number of variables, set as hypotheses, are modeled in the form of linear combinations. Kaplan (2001) defined SEM as "a class of methodologies that seeks to represent hypotheses about the means, variances, and covariances of observed data in terms of a smaller number of 'structural' parameters defined by a hypothesized underlying conceptual or theoretical model". SEM enables analysis that integrates correlation, regression, and factor analysis, and it has advantages in including factors that cannot be directly observed, called latent variables, and sets up multiple dependent variables within a single analysis. Therefore, it is particularly useful for testing the validity of hypotheses involving complex interrelationships of various variables such as a process of creating shared value in urban agriculture.

Part of origins of the method date to the early years of the twentieth century with the development of exploratory factor analysis, usually credited to Charles Spearman. A few years later, the biogeneticist Sewell Wright developed the basics of path analysis. Wright demonstrated how observed covariance could be related to the parameters of a model that represents both direct and indirect causal effects among a set of variables. Wright also invented path diagrams, which are graphical representations of direct and indirect effects that we still use today. The technique of path analysis was subsequently introduced to the behavioral sciences by various authors such as A. Blalock and O. Duncan. The measurement (factor analysis) and structural (path analysis) approaches were basically integrated in the early 1970s in the work of three authors: K. Jöreskog, J. Keesling, and D. Wiley, into a framework that called the JWK model (Kline 2011, p. 15).

Path Diagrams

A path model is a structural model for observed variables, and a structural model represents hypotheses about effect priority. The path diagram in Fig. 3.2 represents the hypothesis by convention, causally prior variables are represented in the left part of the diagram, and their effects are represented in the right part. The line in the figure with the single arrowhead that points from X to Y (1) represents the corresponding direct effect; from X to Z (2) and from Y to X (2). Statistical estimates of direct effects are path coefficients, which are interpreted just as regression coefficients. It is the presence of disturbances (D_1 and D_2) in structural models that signal the assumption of probabilistic causality. Because disturbances can be considered by latent variables in their own right, they are represented with circles (Kline 2011, p. 103).

There are at least four major reasons for the popularity of SEM (Schumacker and Lomax 2010, p. 6). The first reason suggests that researchers are becoming more aware of the need to use multiple observed variables to better understand their

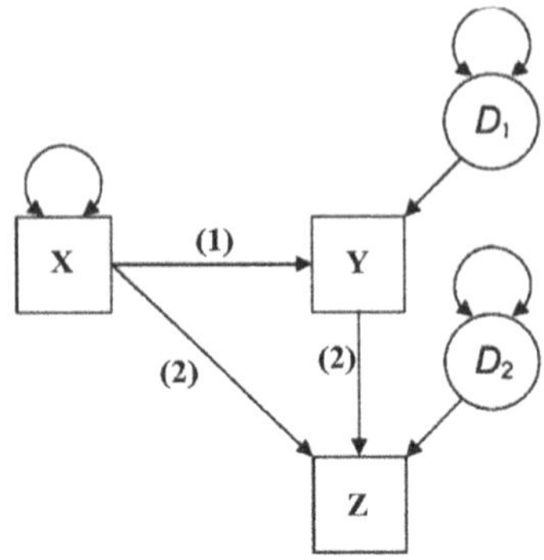

Fig. 3.2 Path diagram of SEM. *Source* Made from Kline (2011, p. 104)

area of scientific inquiry. Basic statistical methods only utilize a limited number of variables, which are not capable of dealing with the sophisticated theories being developed. The use of a small number of variables to understand complex phenomena has the limitation. For instance, the use of simple bivariate correlations is not sufficient for examining a sophisticated theoretical model. In contrast, structural equation modeling permits complex phenomena to be statistically modeled and tested.

A second reason involves the greater recognition given to the validity and reliability of observed scores from measurement instruments. Specifically, measurement error has become a major issue in many disciplines, but measurement error and statistical analysis of data have been treated separately. Structural equation modeling techniques explicitly take measurement error into account when statistically analyzing data.

A third reason pertains to how structural equation modeling has matured over the past 30 years, especially the ability to analyze more advanced theoretical SEM models. For example, group differences in theoretical models can be assessed through multiple-group SEM models. As an example, interaction terms can now be included in an SEM model so that main effects and interaction effects can be tested. These advanced SEM models and techniques have provided many researchers with an increased capability to analyze sophisticated theoretical models of complex phenomena, thus requiring less reliance on basic statistical methods.

Finally, SEM software programs have become increasingly user-friendly. For example, until 1993 LISREL users had to input the program syntax for their models using Greek and matrix notation. Today, most SEM software programs are Windows-based and use pull-down menus or drawing programs to generate the program syntax internally. Therefore, the SEM software programs are now easier to use and contain features similar to other Windows-based software packages.

As mentioned above, SEM is superior to traditional methods of quantitative analysis in the analysis of complex social phenomena. However, Tarka (2018) also appreciates the strengths of SEM in the social sciences but states that SEM should be used appropriately, with an understanding of its technical problems and limitations. In particular, it is important to note that the phenomena related to the change and development of individuals, groups, and societies are all historical phenomena based on the activities of diverse actors. SEM is no different from traditional quantitative analysis in that it relies heavily on correlations among variables, and this reliance on correlations leads to the danger of discarding both the dynamic and qualitative aspects of social phenomena (Sato et al. 2009).

Among these, with regard to the dynamic aspects of social phenomena, SEM has been extended to Dynamic SEM (DSEM) (Asparouhov et al. 2018). However, the data that DESM analyzes are intensive longitudinal data, so it cannot be applied to broad, long-term social phenomena. Therefore, it is important to overcome the above shortcomings by selecting an appropriate qualitative research method and applying mixed methods integrated with SEM. Brown et al. (2021) compared the research approach of SEM and realist inquiry of qualitative research, and examined the effects that integrate the two and pointed out the following challenges of mixed methods.

Since the two methodologies have very different epistemological and procedural characteristics, using them together in mixed methods requires careful translations between their respective paradigm spaces. However, the complementarity of the two approaches may allow each approach to address the limitations of the other and produce significant results. The model of causality in SEM focuses on important relationships that may be missed in the detailed contextual accounts of realist inquiry, and likewise, realist inquiry has the potential to identify specific local mechanisms operating in particular contexts that may be lost in SEM.

3.2.3 Trajectory Equifinality Modeling (TEM)

TEM is one of the qualitative research methods, and it is a methodology of cultural psychology and qualitative psychology (Yasuda and Sato 2012; Sato et al. 2016) for capturing and describing the changes of individual in relation to the society with time being. Therefore, TEM has two features: a systematic approach to treating human as an open system, and an emphasis on the flow of time experienced by the individual.[1]

In this study, the conceptual model of TEM shown in Fig. 3.3 is used. The opportunities of management decision-making (such as business preparation, business start-up, new business development, etc.) were set as the main bifurcation point (BFP). Also, the main socioeconomic environment factors that influence the decision making

[1] The basic concepts of TEM include Irreversible Time, Equifinality Point (EFP), Bifurcation Point (BFP), Trajectory, Social Direction (SD) and Social Guidance (SG). It is thought that there are multiple paths of action and choice of people living in the flow of irreversible time. However, there is no freedom of choice and action, and there is no widespread path. Due to historical, cultural, and socially embedded space–time constraints, there is a point that reaches a certain steady state equally which is equifinality point (EFP). On the other hand, the point at which the path branches off is conceptualized as a bifurcation point (BFP). BFP is not unrelated to irreversible time. The BFP and its branching paths are not inherently fixed in nature as people follow a given path, but occur in a time course that never goes back. Like EFP, it is also a point that occurs in historical, cultural and social influences.

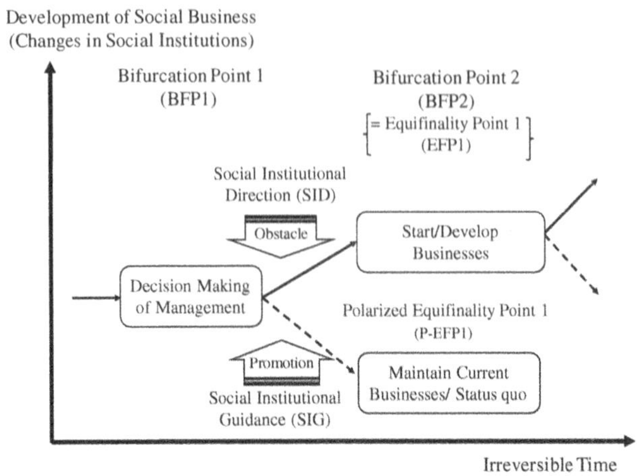

Fig. 3.3 Conceptual model (TEM analysis)

for business start and development are set as social institutional direction (SID) and social institutional guidance (SIG). Since this analysis focuses on the verification of the hypothesis, we will not describe equifinality points (EFP) in detail. Instead, we will analyze the bifurcation point (BFP2) corresponding to the equifinality point (EFP1) as an opportunity for decision making in the next period which is generated by the decision making at the bifurcation point (BFP1) in the previous term. Along with such meaning-based analysis, we will superimpose and organize the aspects of business development as a social business that creates shared value.

The impact of social institutional guidance in the analytical framework on social entrepreneurship (SE) corresponds to SID (social institutional direction) and SIG (social institutional guidance). In addition, the vertical axis in the TEM chart represents the development of social business (SB) and also reflects changes in the social institutions.

References

Asparouhov T, Hamaker EL, Muthén B (2018) Dynamic structural equation models. Struct Equ Model 25(3):359–388

Bamberger M (ed) (2000) Integrating quantitative and qualitative research in development projects. The International Bank for Reconstruction and Development/The World Bank, Washington DC

Brown A, Hecker KG, Bok H, Ellaway RH (2021) Strange bedfellows: exploring methodological intersections between realist inquiry and structural equation modeling. J Mixed Methods Res 15(4):485–506

Creswell JW, Plano Clark VL (2017) Designing and conducting mixed methods research, 3rd edn. Sage, Thousand Oaks

Kaplan D (2001) Structural equation modeling. In: Smelser NJ, Baltes PB (eds) International encyclopedia of social & behavioral sciences, 1st edn. Elsevier, Oxford, pp 15215–15222

Kiminami L, Kiminami A (2006) Sustainability of agriculture and urban quality of life in Japan—economic efficiency, sociality and environment protection. Stud Region Sci 36(2):305–321
Kiminami L, Kiminami A, Furuzawa S (2018) Impacts of multi-functionality of urban agriculture on the CCs in Japan. Asia-Pacific J Region Sci 2(2):507–527
Kline RB (2011) Principles and practice of structural equation modeling, 3rd edn. Guilford Press, New York
Sato T, Hidaka T, Fukuda M (2009) Depicting the dynamics of living the life: the trajectory equifinality model. In: Valsiner J, Molenaar P, Lyra M, Chaudhary N (eds) Dynamic process methodology in the social and developmental sciences. Springer, New York, pp 217–240
Sato T, Mori N, Valsiner J (eds) (2016) Making of the future: the trajectory equifinality approach in cultural psychology. Information Age Publishing, Charlotte
Schumacker RE, Lomax RG (2010) A beginner's guide to structural equation modeling, 3rd edn. Routledge, New York
Tarka P (2018) An overview of structural equation modeling: its beginnings, historical development, usefulness and controversies in the social sciences. Qual Quant Int J Methodol 52:313–354
Tashakkori A, Creswell JW (2007) Editorial: the new era of mixed methods. J Mixed Methods Res 1(3):3–7
Yasuda Y, Sato T (2012) Understanding the life path by using TEM (Trajectory Equifinality Model): new development of qualitative research. Seishin-shobo, Tokyo (in Japanese)

Chapter 4
Urban Agriculture in Japan and China: Trends and Policies

Abstract This chapter explains the trends and policies of urban agriculture in Japan and China by focusing on two megacities of Tokyo and Shanghai. Urban planning in the two cities is compared from the aspects of slogan, vision, approach, and the position of urban agriculture in it. As for the urban agriculture in Japan, the Basic Law for the Promotion of Urban Agriculture was enacted in 2015 with the aim of contributing to the formation of a favorable urban environment through the stable continuation of urban agriculture and the exertion of multiple functions. In recent years, the movement of businesses related to urban agriculture managed by non-farmers/corporate and social enterprises has been increasing, and the cases involving a large number of stakeholders are increasing in Tokyo. On the other hand, the concept of urban agriculture was introduced in China around 1990s, Chinese scholars have studied it for decades combining it with the Chinese context since then. There are three types of urban agriculture in China: small-scale urban agriculture, capital-intensive urban agriculture, and tourism-type urban agriculture (Luehr et al, Thornton A, Urban democracy and governance in north and south. Palgrave Macmillan, 2020). In recent years, urban agriculture policies in Shanghai have been planned and implemented with an emphasis on demonstrating multi-functionality of agriculture.

Keywords Urban agriculture · Japan · China · Urban planning · Policy

4.1 Introduction

While Japan is currently experiencing a decline in the population overall, Tokyo is an exception and keeps growing, boasting the largest number (more than 38 million) of people including the adjacent cities of Chiba, Kanagawa and Saitama. As the population concentrates in Tokyo more and more, rural areas and small cities suffer from a severe decline. On the other hand, there is a new trend of starting agricultural-related business in Tokyo (Ono 2018) in recent years, as well as the migration to local areas from large cities.

On the other hand, as China's urbanization progressed, the demands for foods such as fruits, vegetables, and animal products, as well as for a clean, safe and

L. Kiminami et al., *Social Entrepreneurship, Social Business and the Multi-functionality of Urban Agriculture*, SpringerBriefs in Economics,
https://doi.org/10.1007/978-981-19-1762-2_4

healthy living environment are rising. For example, the result of survey targeting Shanghai residents showed that the residents place great hope on building Shanghai into an "environment-friendly, economic developed, culturally diversified, and safe and livable city" (Shanghai Master Plan 2017–2035). This means that the 'soft' factors of preferences for openness, tolerance and amenity will drive the location choice of 'creative classes', and the expansion of the urban middle class will create new demands for urban agriculture to meet their needs for quality of life. In other words, in addition to the traditional demand for agricultural products, there are ecological and other demands, and people can afford to pay for them now (Shi et al. 2011). Under such circumstance, new types of agri-business such as tourist farms, farm restaurants and educational farms have emerged in some peri-urban areas to meet the demands of urban residents for agriculture amenity (Peng et al. 2015). Furthermore, community gardens and rooftop gardens are also appearing in the center area of large cities, where urban residents have developed new relationships with their communities through the experience of urban agriculture, increasing satisfaction with their local life (Ge et al. 2021).

The purpose of this chapter is to explain the current situations and policies of urban agriculture in Japan and China by focusing on the megacities of Tokyo and Shanghai in the two countries.

4.2 Urban Planning and Urban Agriculture in Tokyo and Shanghai

Table 4.1 shows a comparative summary of the city plans of Tokyo and Shanghai based on their vision, goals, main contents, and the position of urban agriculture. It is clear that both cities are aiming to make a sustainable development not only focusing on economic aspect but also on residents' quality of life.

In the city plan of Tokyo, conservation and utilization of urban farmland are positioned in one of the seven strategies (Strategy 6: Green and Water Environment). In addition to the development of agriculture as an industry, multi-functionality of agriculture is also clearly positioned to be promoted in the plan. On the other hand, in the city plan of Shanghai, the need for harmony between ecological space, agricultural space, and urban space is stated and urban agriculture is positioned to alleviate the negative aspects of urban land use.

4.3 Urban Agriculture in Japan

In Japan, after the late 1950s, as the urbanization progressed, green areas in built-up urban areas are disappeared, urban areas are expanded at outskirts, and agricultural and forest lands are destroyed. Unorganized progress of turning lands into housing

Table 4.1 Comprehensive city plans in Tokyo and Shanghai

	Tokyo	Shanghai
Slogan	"*A Highly developed mature city* that is relaxing and full of vitality"	"*Better City, Better Life*"
Vision	A city selected by people around the world A zero emission city A sustainable city	An excellent global city. An admirable city of innovation, humanity and sustainability A modern socialist international metropolis with world influence
Approach	Seven strategies, 30 policies S1: creating the hubs full of vitality that generate sustainable growth S2: achieving flexible, unrestricted exchange of people, goods, and information S3: building a city that addresses the risk of natural disasters and environmental issues S4: building a supportive and comfortable environment for all S5: realizing highly convenient lifestyles and creating diverse communities S6: building a city with a network of beautiful seasonal greenery and water S7: creating new appeal through the arts, culture, and sports	Three goals: more dynamic: an innovation city more attractive: a humanistic city more sustainable: an eco-city New model based on the 3 approaches (bottom-line control/endogenous development/flexible adaptation)
Urban agriculture	**Nurture urban agriculture** which plays a role in industry and invigorates the community 1. Farm stand where consumers can interact with farmers 2. Vegetable factory that incorporates solar power generation and is self-sufficient energy-wise 3. "Smart agriculture" that leverages advanced technologies to efficiently produce high-quality products 4. Invaluable rice paddies that provide opportunities to experience planting and harvesting 5. Restaurant established adjacent to a farm	Coordinate and optimize **"three spaces"**, i.e., ecological, agricultural and urban spaces, in the city; promote integrated utilization of space; establish a **"four lines"** control system covering ecological conservation red line, permanent prime cropland conservation red line, urban development boundary and cultural protection control line
Source	"Grand Design of Urban Development" (http://www.toshiseibi.metro.tokyo.jp/keikaku_chousa_singikai/grand_design.html) Target year: 2040	Shanghai Master Plan 2017–2035 (http://www.shanghai.gov.cn/newshanghai/xxgkfj/2035004.pdf) Target year: 2035

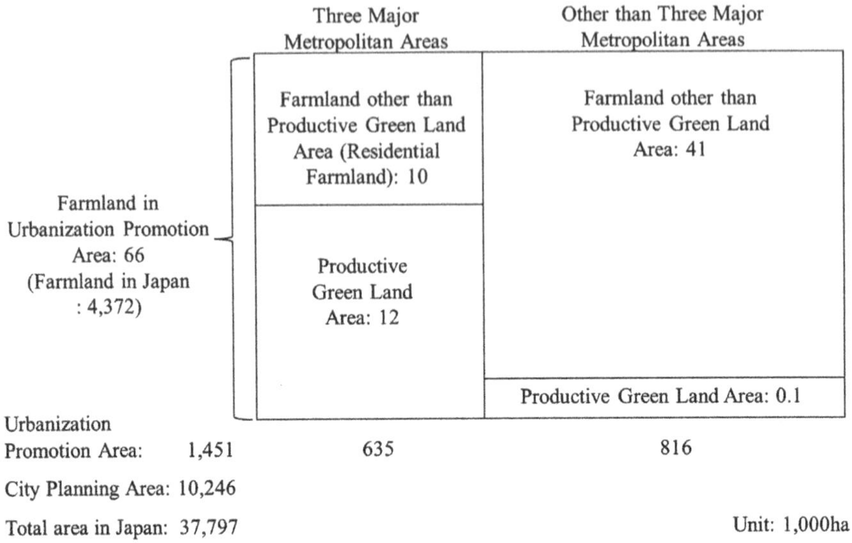

Fig. 4.1 Farmland in the urbanization promotion area of three major metropolitan areas (2020). *Source* Ministry of Land, Infrastructure, Transport and Tourism "Current Status of Farmland Area"

lands from the urban core to outskirts without sufficient urban infrastructure resulted in a so-called "sprawl phenomenon". On the contrary, farmlands were brought into urban areas in an expectation of turning them into housing lands in the future. As a result, urban areas are now dotted with farmlands. These remaining farmlands in urban areas are lands suitable for agriculture and in many cases, farmers have excellent agricultural management abilities. In addition, these remaining farmlands are functioning as green spaces in place of urban parks.

For a systematic improvement of urban areas, City Planning Areas are designated and they are classified into urbanization promotion areas (UPAs) and urbanization control areas (UCAs). UPAs include existing residential areas and areas that would be developed within a decade. Farmlands located in UPAs can be used for non-agricultural purpose without the need for permission, by simply submitting a notice to the government. However, in the case of UCAs, land use was strictly regulated and permission for land conversion was required. Nowadays, farmlands in the UPAs in three major metropolitan areas[1] are classified into Productive Green Lands (farmlands to be preserved) and Residential Farmlands (farmlands to be turned into housing lands) (see Fig. 4.1). Productive Green Lands refer to farmlands that are designated as "lands with functions to prevent pollution or disaster and to provide good living environment such as preserving urban environment that are agriculture and fishing friendly and lands that are suitable for building public facilities". If farmland is designated as Productive Green Land, the designation cannot be canceled for a period of

[1] Three Major Metropolitan Areas consist of the Tokyo metropolitan area, the Chukyo metropolitan area, and the Kansai metropolitan area.

thirty years and the land is required to be used for agriculture. The property tax rate for such lands is the same as that for farmlands and deferred payment of inheritance tax applies to such lands. On the contrary, the property tax rate for Residential Farmlands is the same as that of residential lands but much higher than that of normal farmlands, and the deferred payment for inheritance tax is not accepted.

The Basic Law for the Promotion of Urban Agriculture was enacted in 2015 with the aim of contributing to the formation of a favorable urban environment through the stable continuation of urban agriculture and the exertion of multiple functions. The principles of the Urban Agriculture Promotion Basic Law are as follows: (1) Appropriate and sufficient exertion of various functions of urban agriculture, effective utilization and proper conservation of urban farmland; and (2) Coexistence with agriculture should be achieved in good urban formation; and (3) Measures should be promoted with the understanding of the people, the government is required to take necessary legal, tax, and financial policies. Furthermore, the government is obliged to formulate an urban agriculture promotion basic plan so that measures can be promoted comprehensively and systematically.

According to Article 2 of the Basic Law for the Promotion of Urban Agriculture, urban agriculture is defined as "agriculture carried out in and around urban areas.“ Therefore, it is not easy to specify the area range, but in relation to urban planning, agriculture in "urban areas" is often regarded as urban agriculture because of its institutional position such as taxation. An urbanized area is an "area that has already formed an urban area and an area that should be prioritized and systematically urbanized within approximately 10 years."

Comparing agriculture in urbanized areas with that of Japan as a whole, the share of the number of agricultural management entities is 10.8 (MAFF 2021). However, since the share of agricultural land in the urbanized area is only 1.5%, the scale of cultivated land for agricultural management in the urbanized area is small (See Table 4.2). On the other hand, agriculture in urbanized areas is carried out by taking advantage of the location characteristics of being close to a large number of consumers, the share of agricultural output value in the whole country accounts for 7.0%. Furthermore, looking at the characteristics of urban agriculture from the sectoral composition of agricultural products, the proportion of vegetables is extremely high (See Table 4.3).

Table 4.2 Share of agriculture in urbanization promotion area in 2020

	Agricultural management entity (1000)	Farmland (1000 ha)	Agricultural production (billion yen)
All areas: a	1076	4372	8894
Urbanization promotion area: b	140	64	6229
b/a (%)	10.8	1.5	7.0

Source MAFF(2021)

Table 4.3 Percentages of agricultural output by sector in major cities in 2019 (Unit: %)

	Rice	Potatoes	Vegetables	Fruits	Flowering plants	Livestock
Japan	19.2	2.2	25.6	9.3	3.7	35.5
Tokyo special ward	0.0	3.7	69.3	6.7	12.5	6.9
Yokohama city	1.2	3.9	61.2	8.2	10.9	12.5
Nagoya city	20.7	1.7	51.2	12.8	N.A.	7.4
Osaka city	7.8	0.0	78.4	0.0	5.9	7.8

Source MAFF Statistics of Agricultural Income Produced 2019, Agricultural output by municipality (estimate) 2019

Furthermore, the Law on Green Production Land has been amended in 2017 and the Law Concerning the Facilitation of Leasing of Urban Farmland has been enacted in 2018 in order for the development of business and the fulfillment of multiple functions of urban agriculture. Urban agriculture includes different forms of business, such as experience farms/allotment gardens, off-market distribution such as direct sales and school meals, disaster prevention cooperation farmland, green areas/waterside spaces in urban areas, etc., and many of them are considered to have the function of positive externalities. Furthermore, new movements such as plant factories and rooftop farms that are not dependent on urban farmland are appearing.

On the other hand, according to the results of a questionnaire survey[2] conducted by the authors on metropolitan residents, the social problems in metropolitan cities are raised as follows: 'Weakening of local communities'; 'Deterioration of natural environment'; 'Lack of measures against natural disasters'; 'Decline of local traditions and culture'; 'Lack of employment opportunities' etc. (See Fig. 4.2). Although, multi-functionality of urban agriculture is evaluated by about 70% of the residents only about 10% of them have actually experienced urban agriculture (See Fig. 4.3).

4.4 Urban Agriculture in China

China's cities have enjoyed several decades of explosive growth. The urbanization rate had already exceeded 59.2% by 2018 (UN 2019). However, international comparisons suggest that when countries pass a level of urbanization of around 50%, the relationship between urbanization and economic growth depends critically on how urbanization processes are managed.

Around 1990s, the concept of urban agriculture was introduced in China, and since then, Chinese scholars have studied it for decades, combining it with the Chinese context (Fang et al. 2008). However, the academic origin of urban agriculture in China is not from the West, but is from Japan (Yu et al. 1998). With the influences from Japan, food security, urban poverty alleviation, economic benefits, sustainability, and

[2] The web questionnaire survey was conducted from March 10 to March 12, 2021.

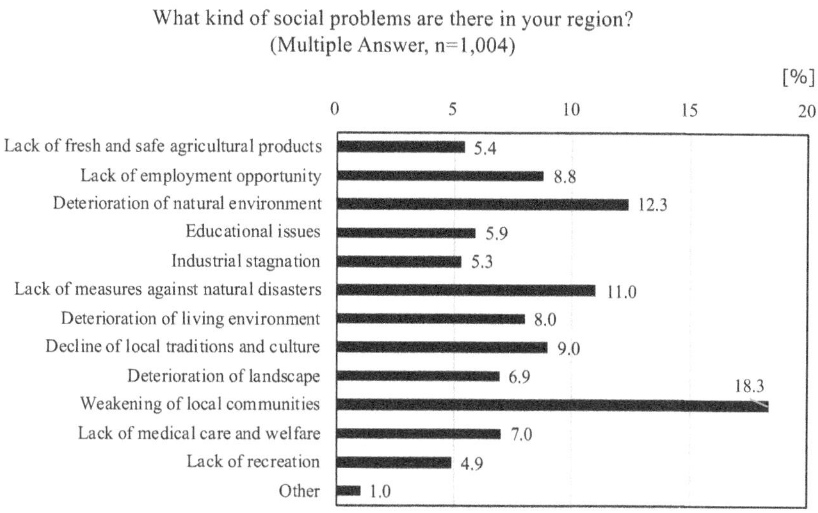

Fig. 4.2 Social problems in urban areas in Japan. *Source* Authors' Original Questionnaire Survey

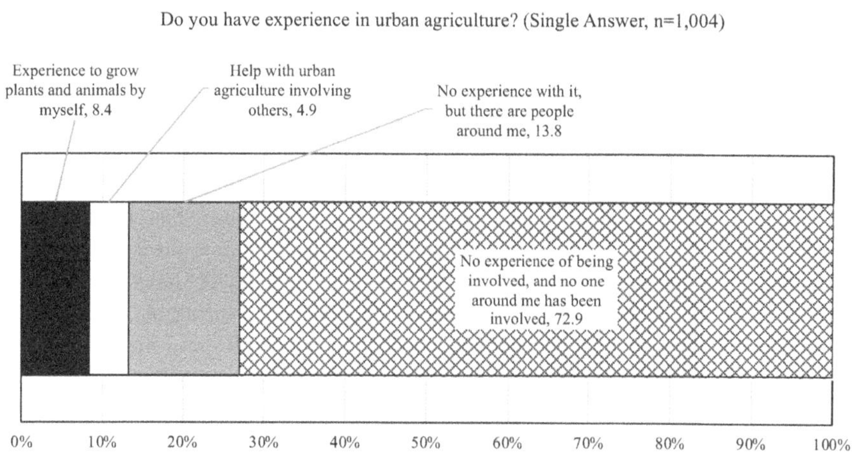

Fig. 4.3 Involvement of urban agriculture by urban residents in Japan. *Source* Authors' Original Questionnaire Survey

livability are the essential points of the Chinese urban agriculture promotion. Since urban agriculture is a relatively new concept for Chinese society and planning, the amount of available literature and other forms of data about urban agriculture in China is limited. In the following, we will summarize the characteristics of urban agriculture in China according to Zhu (2016).

Firstly, urban agriculture in China is focused on economic and temporal perspectives, especially on ‘modern’ aspects (the supports from agricultural facilities, mechanization, and high technology to increase productivity of urban agriculture), instead of defining it mainly from spatial perspective. Secondly, the Chinese urban agriculture practices mainly happen in suburban instead of inner-urban area, and the government or authorities’ attitudes on inner-urban agriculture in China are considerably neutral, without supporting nor opposing. Thirdly, urban agriculture contributes to accelerating the urban–rural integration in China via decreasing the income and life quality disparity between urban and rural residents (creating employment especially for rural immigrants in cities). Finally, the target group of urban agriculture is centered on the middle class and the wealthy class.

According to Luehr et al. (2020), there are three types of urban agriculture in China: small-scale urban agriculture, capital-intensive urban agriculture, and tourism-type urban agriculture. Small-scale urban agriculture tends to be run by residents, capital-intensive and tourism-based urban agriculture tends to be run by private companies and the tourism type tends to be located in the suburbs.

In recent years, urban agriculture policies in Shanghai have been planned and implemented with an emphasis on demonstrating multiple functions. In 2011, the “12th Five-Year Plan for Modern Agricultural Industry of Shanghai” was enacted, and it is proposed to significantly enhance the economic, ecological and service functions of urban agriculture. Regarding agricultural production, it emphasized that quantitative protection and quality improvement of agricultural land are the most important. And “cultivation of agricultural companies”, “construction of farms for high quality foods and vegetables”, “strengthening of agricultural quality certification” and “cultivation of agricultural product brands” are emphasized as well.

Since the “13th Five-Year Plan” period, Shanghai’s agriculture has generally been in a development stage where both quantity decline and quality improvement coexist (See Table 4.4). In the period of 2015–2019, the overall scale of Shanghai’s agricultural industry is shrinking, which is mainly reflected in the shrinking scale of agricultural production such as the decrease in sown area, poultry, pig meat, vegetable and grain production, except raw milk. On the other hand, with the reform of the agricultural supply-side structure and the implementation of the rural revitalization strategy, Shanghai’s agricultural structure adjustment has undergone profound changes. Compared with 2015, the share of animal husbandry has fallen by more than 7%, the share of planting has fallen by about 0.1%, and the share of forestry, fisheries, and agriculture, forestry, animal husbandry and fishery services have all increased by more than 2%. In addition, the integration of agriculture and non-agricultural industries accelerated the development process, agricultural modernization has reached a new level and agricultural green development is at the forefront of the country. Therefore, with the changes in internal and external environmental conditions and development stage, it is great significance to rethink the policy orientation of agriculture including multi-functionality of urban agriculture in the overall development of Shanghai during the period of “14th Five-Year Plan”.

By comparing the urban agriculture of Japan’s big cities and Shanghai, the following differences and similarities are clarified by Fang (2019). First, Japan has

Table 4.4 Achievement of "13th five-year plan" of Shanghai agriculture

		Achievement
Agricultural production	Production amount (2015–2019)	Sown area—26.3%; poultry—56.6%; pig meat—41.5%; vegetable—25.8; grain—14.4%; raw milk—7.4%
	Production value (2015–2019)	Primary industry—17.2%; production value of agricultural industry 28.08 billion yuan (2019)
Structural adjustment of agriculture	Agricultural industry structure (2019)	Ratio of production value: agriculture 51.6%, forestry 6.4%, livestock 17.3%, fishery 19.2%
	Integration of agriculture and non-agricultural industries	Implementation of the rural revitalization strategy: agricultural and tourism, cultural and creative industries etc. in the suburbs of Shanghai are showing a strong momentum of development
	Changes in the production structure	In 2018, the total agricultural output value of Shanghai's outside-owned farms increased by 25.6%. The main products are vegetables, pig meat, milk, and aquatic products Increase the area of green manure and deep plowing and drying, and complete the "three zones": delineation of the functional area of food production, the protection area of vegetable production and the advantage area of characteristic agricultural products
Innovation and transformation	Agricultural modernization	Shanghai was approved to establish a national modern agriculture demonstration zone (2015) All 9 agricultural-related districts in the city met the certification standards (2019) Jinshan Modern Agricultural Industrial Park was selected as the national modern agricultural industrial park (2019) Modern agricultural enterprises, cooperatives and farms by the end of 2019: 269 key leading agricultural industrialization enterprises (24 national-level; 88 municipal-level), 2757 farmers' professional cooperatives (86 national-level; 124 municipal-level), 4347 family farms (76 municipal-level)
	Agricultural green development	Chongming District was listed as the first batch national green development pioneer areas in 2017, and was selected as one of the top ten models in 2019 Songjiang District was selected as the second batch of national agricultural green development pioneer areas (2019) The city's green food certification rate reached 20%, the use of chemical fertilizers and pesticides has dropped by 26.7 and 34.6% in 2019 The utilization rate of livestock and poultry breeding waste resources reached 96%, and the supporting rate of manure treatment facilities was 100% in 2019

(continued)

Table 4.4 (continued)

		Achievement
	Promotion of agricultural branding	Incorporating agricultural product brand building into the assessment of rural revitalization, and identifying 20 real estate agricultural product brands in 9 agricultural-related areas Songjiang rice, chongming rice, nanhui peach, and malu grape were selected in the national agricultural brand catalog 2019 as a regional public brand 14 agricultural products in the city have been awarded National Geographical Indications by 2019

Source Made fromXiu (2020)

formulated a number of laws on the development of urban agriculture at the national level for a long time, and the local governments have formulated detailed and operable laws and regulations. Although Shanghai has formulated some local laws and regulations to support the development of urban agriculture and issued many professional planning and support policies in recent years, the instability and arbitrariness of policies, and the conflicts of policy documents have occurred from time to time. Secondly, against the trend of the aging of farmers, Japan formulated guiding policies to attract the "three haves" (aspirational, intelligent, and resourceful) personnel engaged in agriculture, cultivated various new types of agricultural entities and built a good atmosphere. In contrast, Shanghai has some shortcomings in cultivating new agricultural entities. At present, new professional farmers in Shanghai are limited to the children of farmers with local household registration.

Thirdly, Japan's urban agriculture has a large number of preferential policies, with various policy subsidies accounting for 40% to 50% of farmers' income. At present, Shanghai's agricultural support policy measures are in line with Japan's policies, the level of various financial subsidies is the highest in the country, and financial income accounts for 22% of the total income of farmers. Due to the government's excessive protection, Japanese agriculture is not competitive in the international arena and it is difficult to sustainably develop. Therefore, it is necessary to step up the adjustment and improvement of the support forms and operating methods of Shanghai's agricultural policy. The policy orientation must focus on promoting green production methods and high quality of products, and improve the efficiency of financial subsidies.

However, when comparing multiple perspectives of social entrepreneur, social business and cities, the characteristics of urban agriculture in Tokyo and Shanghai can be shown in Table 4.5. In Tokyo, agricultural land and residential areas are coexisting, and there are many farmers in the city center. The percentage of social entrepreneurs from farmers who work on social business, which exerts multiple functions of agriculture, is relatively high, but the number of companies entering the market was small. Therefore, the ability to involve stakeholders in urban agriculture is relatively weak. However, in recent years, the movement of businesses related to urban agriculture managed by non-farmers/corporate and social enterprises has been

Table 4.5 Comparison of Tokyo and Shanghai

	Tokyo	Shanghai
Distance between urban agriculture and residents	Close	Far
Farmland in the city	Few	Very few
Traditional farming in urban farming	Many	Almost none
Citizens' interest in social enterprises	High	Low
Social entrepreneur in urban agriculture	Farmer/From non-farmer	From non-farmer/enterprises

increasing, and the cases involving a large number of stakeholders are increasing in Tokyo.[3] On the other hand, in Shanghai, since there is no agricultural land in the center of the city, urban residents and companies are developing the social businesses that demonstrate the multi-functionality of agriculture involving stakeholders while utilizing non-farmlands.

References

Fang Z, Wu F, Wang W (2008) A literature review of disputes and issues in the research on Chinese urban agriculture theories. Chin Agric Sci Bull 24(8):521–525 (in Chinese)

Fang Z (2019) Urban agricultural construction in big cities of Japan and its enlightenment to the positioning of Shanghai's agricultural development. Sci Devel (kexue Fazhan) 128:95–102 (in Chinese)

Ge Y, Kiminami L, Furuzawa S (2021) Perception and behavior of residents towards community gardens and satisfaction of local life: empirical analysis on the case of Shanghai, China. Asia-Pacific J Region Sci. https://doi.org/10.1007/s41685-021-00219-2

Luehr G, Glaros A, Si Z, Scott S (2020) Urban agriculture in Chinese cities: practices, motivations and challenges. Thornton A. Urban democracy and governance in north and south. Palgrave Macmillan, London, pp 291–309

MAFF (2021) Situation about the urban agriculture. https://www.maff.go.jp/j/nousin/kouryu/tosi_nougyo/attach/pdf/t_kuwashiku-17.pdf

Ono A (2018) Tokyo agricultural creators. Ikaros Publication Ltd., Tokyo (in Japanese)

Peng J, Liu Z, Liu Y, Hu X, Wang A (2015) Multifunctionality assessment of urban agriculture in Beijing City, China. Sci Total Environ 537(15):343–351

Shi Y, Cheng CW, Lei P, Zhu Y, Jia , Wen TJ (2011) Analysis on the relationship between urban eco-agriculture development and emergence of middle income class of cities. Guizhou Social Sci (02):55–60 (in Chinese)

UN (2019) World urbanization prospects. The 2018 revision, United Nations, New York

Xiu B (2020) Review of the '13th five-year plan' and thinking of the '14th five-year' of shanghai urban agriculture development. Shanghai Rural Econ 2020(12):13–17 (in Chinese)

[3] See https://tokyourbanfarming.jp/; https://www.shibuya-univ.net/professor/detail/1279/ for details.

Yu J, Zhang Z, Bai E, Cheng Z, Wang C, Xu X, Pan Y, Wang S (1998) Preliminary study on origin and definition of the term "city agriculture": investigation on Japan city agriculture theory. Acta Agric Shanghai 14(2):79–84 (in Chinese)

Zhu X (2016) Planning for urban agriculture in China: an exploratory research on urban agriculture from the Chinese planning perspective. MSc thesis of the master urban environmental management, Wageningen University

Chapter 5
Creative Classes, Social Business and the Multi-functionality of Urban Agriculture in Tokyo and Shanghai

Abstract The purpose of this chapter is to clarify the relationships among creative classes (CCs), social business (SB) and the multi-functionality of urban agriculture (MFUA) in Tokyo and Shanghai, the two global cities of Asia. Specifically, structural equation modeling (SEM) is introduced to the results of authors' original questionnaire survey targeting the residents in the two cities. As for the creative class (CC) in this chapter, we also pay attention to the people of creative thinking besides of creative occupations. Their orientation toward multi-functionality of agriculture and their relationships with social capital and social business are clarified. Based on the analytical results, we draw policy implications on the participation of CCs (both creative occupation and creative thinking) in social business for solving social problems by demonstrating the multi-functionality of urban agriculture in global cities.

Keywords Multi-functionality of urban agriculture (MFUA) · Creative class (CC) · Structural equation modeling (SEM) · Tokyo · Shanghai

5.1 Introduction

The main targets of this study are the two megacities of Tokyo in Japan and Shanghai in China, with the highest population in East Asia. Table 5.1 summarized the situation in the two cities in terms of the roles of urban economy, creative class, and agricultural sector.

First, the two cities are not only large in the size of population but also in economic activities, and are growing continuously. In particular, Shanghai had a high population growth rate at the average of 2.6% over the past 15 years, and a high growth rate of gross regional product (GRP) per capita under the rapid urbanization. In other words, the economies of agglomeration created the increase in urban productivity and population at the same time.

This chapter is a revision based on (Kiminami et al. 2020).

L. Kiminami et al., *Social Entrepreneurship, Social Business and the Multi-functionality of Urban Agriculture*, SpringerBriefs in Economics,
https://doi.org/10.1007/978-981-19-1762-2_5

Table 5.1 Basic situations in Tokyo and Shanghai

		Unit	Tokyo	Shanghai
Urban economy	GRP per capita (2015)	US$/person	59,518	15,655
	Annual growth rate of GRP per capita (2000–15)	%	0.2	10.5
	Population density (2015)	person/km^2	6169	3809
	Annual growth rate of population (2000–15)	%	0.8	2.6
Creative class	Ratio of population engaged in CC (2010)	%	16.9	15.0
	Annual growth rate of population engaged in CC (2000–10)	%	0.3	6.1
Gender gap of CC	Male CC/female CC (2010)		1.4	1.0
	Male CC/female CC (2000)		1.6	0.8
Foreigner	% of foreigner (2015)	%	3.1	0.7
Environmental amenity	Park area per person (2015)	m^2/person	5.7	1.0
Agricultural sector	Ratio of population engaged in agric. (2015)	%	0.4	3.6
	Ratio of agric. in GRP (2015)	%	0.1	0.4
	Agric. GRP per person engaged in agric., fores. and fishery (2015)	US$/person	20,838	3,371
	Annual growth rate of agric. GRP per agric. capita (2000–15)	%	2.3	8.5
	Ratio of farmland (2015)	%	3.8	29.9
	Change rate of farmland (2000–15)	%	−20.2	−39.8

Source Tokyo "Tokyo Statistical Yearbook 2015", "Census (2010, 2015)"
Shanghai City "Shanghai Statistics Yearbook 2016", "China Statistics Yearbook (2004, 2016)", "China Population Census (2000, 2010)"
Note The occupational classification code is as follows: For Tokyo, professional jobs (7), technical jobs (8), and arts (writers, artists, entertainers) jobs (9) in the socio-economic classification (22 categories) in the national census. For Shanghai City, scientific researchers (206), industrial engineers (207), agricultural engineers (208), airplane and ship engineers (209), and sanitary engineers (210) in the middle classification (206-219), commercial expert (211), financial expert (212), legal expert (213), education expert (214), literary artist (215), sports expert (216), writer/reporter/editor (217), the total working population of religions (218) and other professionals (219). Therefore, it is necessary to pay attention to the above differences when comparing the two cities strictly

Next, considering comparability, we set the creative class as professional/technical workers and look at its rate of change. The rate of CC in the population in Tokyo is slightly higher (16.9% in Tokyo and 15.0% in Shanghai), but Shanghai has a higher growth rate of CC (0.3% in Tokyo and 6.1% in Shanghai).

Moreover, regarding the role of the agricultural sector in the two cities, the proportion of the agricultural sector in the total economy is small in terms of added value and employment. In addition, the agricultural GRP per capita is significantly lower than the GRP per capita in each city (about one-third in Tokyo and one-fifth in Shanghai). However, productivity in the agricultural sector has risen over the last 15 years (2.3% in Tokyo and 8.5% in Shanghai) when looking at changes in agricultural GRP per capita.

Finally, farmland contributes greatly to the land use in Shanghai occupying about 30% of the city area, but farmland is only about 4% of the city area in Tokyo. Moreover, farmland had decreased by about 20% in Tokyo and by about 40% in Shanghai over the past 15 years. On the other hand, the park area is about 6m^2 per person in Tokyo, while it is about 1m^2 per person in Shanghai.

The purpose of this chapter is to clarify the relationships among creative classes (CCs), social business (SB) and multi-functionality of urban agriculture (MFUA) in Tokyo and Shanghai, the two global cities of Asia. Specifically, based on the previous studies on creative classes (CCs), multi-functionality of agriculture (MFA), social capital (SC) and social business (SB) referred in Chap. 2, we construct a model showing that these factors contribute to the sustainability of global cities (see Fig. 5.1), and verify the model by introducing structural equation modeling (SEM) to the results of authors' original questionnaire survey targeting the residents in the two cities. In addition to creative occupations, we also pay attention to the people of creative thinking, as well as their orientation toward multi-functionality of agriculture and their relationships with social capital and social business.

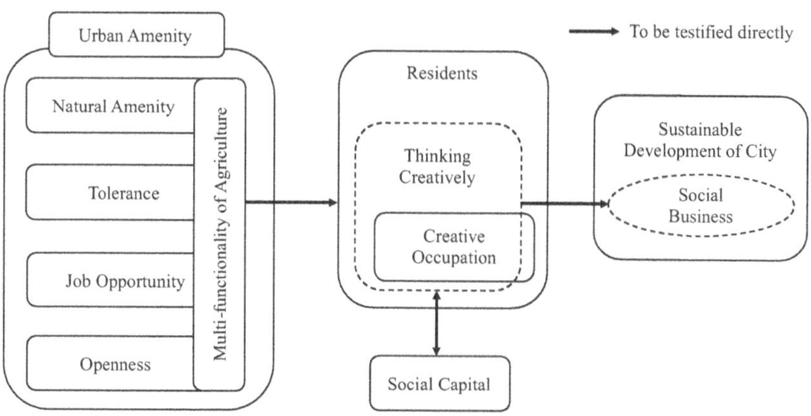

Fig. 5.1 The relationship between multi-functionality of urban agriculture (MFUA) as urban amenity and creative classes (CCs). *Source* Revised from Kiminami et al. (2018, Fig. 3)

5.2 Data and Analytical Method

5.2.1 Data

In this study, we introduce structural equation modeling (SEM) to the results of the questionnaire survey targeting the residents of Shanghai and Tokyo in line with the above framework. The questionnaire survey was conducted through Macromill, Inc. using the method of Internet research. Both surveys ended when the number of respondents reached 1,029 samples. In addition, to clarify the different influences between creative occupation and non-creative occupation in the results of answer, the conditions of assignment including "managers and executives", "professionals/technical workers", and "company officers who are neither manager nor professional or technical work" were evenly distributed. The survey dates were September 22–24, 2017 for Tokyo and January 24–27, 2018 for Shanghai. The questionnaires for Tokyo and Shanghai had the same content, but Japanese and Chinese were used for each.

Here, for the questions of creative thinking, we adopted the definition by the O * NET content model,[1] following the method of McGranhan and Wojan (2007). Specifically, as for the question "How important is 'creative thinking' to the outcome of your current work? (If you work for multiple companies, please answer mainly about the company you work for)" we set the answer items based on the Likert scale (5 levels). In addition, "Creative thinking is related to new uses of ideas, relationships, system or product development, design and production (including art)" is added as a supplementary explanation.

Table 5.2 summarized the results of simple tabulation of respondents' profiles and creative thinking. Compared to the populations of both cities, the results of the questionnaire showed gender and age bias in Tokyo and age bias in Shanghai (see note in Table 5.2 for details). By setting age and gender as control variables in the structural equation modeling, it is considered that the influence of these biases can be eliminated to some extent. However, the simple tabulation results for each question item cannot be interpreted as representing the level of each region as a whole.

Table 5.3 summarized the results of cross tabulation between creative thinking and occupation. The average scores are highest in the order of "managers and executives", "professionals/technical workers", and "others" both in Tokyo and Shanghai. However, even in the "others" there is a certain percentage of respondents who answered that creative thinking is important. Therefore, it is considered that creative thinking is also important in other occupations although there is a strong positive relationship between creative occupation and creative thinking. In the structural equation modeling in the next section, the index of importance of creative thinking will be used as a variable.

[1] Please see O*NET Content Model (https://www.onetcenter.org/content.html) for more details.

Table 5.2 Profile of respondents

Items		Regions	Distribution(%)				
Socio-economic attribute	Sex		Male	Female			
		Tokyo	76.3 (51.3)	23.7 (48.7)			
		Shanghai	50.7 (57.1)	49.3 (42.9)			
	Age		18–19	20–29	30–39	40–49	
		Tokyo	–	2.1 (14.5)	12.4 (18.0)	30.8 (19.7)	
		Shanghai	1.1 (–)	18.0 (26.1)	47.1 (20.3)	18.3 (18.5)	
			50–59	60–69	70 and above		
		Tokyo	32.7 (14.1)	18.1 (14.4)	3.9 (19.3)		
		Shanghai	9.6 (17.7)	5.7 (9.1)	0.2 (8.4)		
	Household income	Tokyo (Million Yen/Year)	Less than 200	200–400	400–600	600–800	800–1000
			0.8	11.6	17.5	14.5	13.4
			1000–1200	1200–1500	1500–2000	2000 and above	I don't know/non-answer
			8.5	8.6	5.6	5.0	5.2
		Shanghai (Yuan/Month)	Less than 10,000	10,000–11,999	12,000–13,999	14,000–15,999	16,000–17,999
			1.5	2.7	4.7	7.1	7.4

(continued)

Table 5.2 (continued)

Items		Regions	Distribution(%)				
			18,000–19,999	20,000–29,999	30,000–39,999	40,000 and above	I don't know
			12.6	25.8	19.7	18.5	0.1
Farmland			There is a lot	There is a bit	There is not much	There is not any	
		Tokyo	6.9	29.2	32.9	31.0	
		Shanghai	12.0	40.5	36.2	11.4	
Creative class	Occupation		Executive in a company	Professional/technical worker	The other worker		
		Tokyo	33.3	33.3	33.3		
		Shanghai	33.3	33.3	33.3		
	Thinking creatively		Extremely important	Very important	Important	Somewhat important	Not important
		Tokyo	18.7	24.5	22.5	18.5	15.8
		Shanghai	38.9	44.2	14.0	2.7	0.2

Note 1 Regarding the age group classification, the registered data of the respondents was used for the ages of the respondents in Tokyo

Note 2 The figures in parentheses are the values of "Labor Force Survey (Tokyo)" for sex in Tokyo, "Census 2015" for age, and "Population Census 2010" for Shanghai

Note 3 For the distribution of respondents by sex and age, an independence test (the null hypothesis is that the distribution of respondents is the same as the distribution of the population) was performed. As a result, the null hypothesis was rejected in Tokyo for sex and age, and in Shanghai for age. For the distribution of the population, the statistical values of "Census 2015" in Tokyo and "Population Census 2010" in Shanghai were used

[Tokyo] Sex: Chi-square statistic 155.0 (p-value = 0.000); Age: Chi-square statistic 575.0 (p-value = 0.000)

[Shanghai] Sex: Chi-square statistic 0.148 (p-value = 0.701); Age: Chi-square statistic 527.5 (p-value = 0.000)

Table 5.3 Cross tabulation about occupational CC and thinking creatively

			Tokyo				Shanghai		
			Executives in a company	Professional/technical worker	Non-professional/technical worker		Executives in a company	Professional/technical worker	Non-professional/technical worker
Thinking creatively	(4)	18.7	29.7	17.5	8.7	38.9	50.1	40.2	26.2
	(3)	24.5	27.7	29.4	16.3	44.2	38.8	46.6	47.2
	(2)	22.5	21.0	25.1	21.6	14.0	9.0	10.8	22.2
	(1)	18.5	13.4	16.3	25.7	2.7	2.0	2.0	4.1
	(0)	15.8	8.2	11.7	27.7	0.2	0.0	0.3	0.3
Average score		2.12	2.57	2.25	1.53	3.19	3.37	3.24	2.95

Note Thinking creatively gives the following scores respectively, and calculated average score. Extremely important (4), Very important (3), Important (2), Somewhat important (1), Not important (0)

5.2.2 Variable Setting and Analytical Method

Table 5.4 summarized the results of cross tabulation and variable settings of the multi-functionality of urban agriculture and creative thinking. Items related to the multi-functionality of urban agriculture were set based on previous studies (OECD 2001; Kiminami and Kiminami 2006, 2007) and related survey results (the Tokyo Metropolitan Government's Bureau of Industry and Labor in recent years).

Specifically, regarding the evaluation of urban agriculture, this is the result of asking the question "Urban agriculture has various roles for life in the city. Please select all of the following roles of urban agriculture that you think are important (Multiple selections are possible)". A supplementary explanation "Urban agriculture is an activity such as crop cultivation and animal breeding carried out in urban areas and surrounding areas" (FAO[2]) was given in the questionnaire.

First, both in Tokyo and Shanghai, the level of creative thinking is positively correlated to the percentage of respondents who answered the multi-functionality of urban agriculture is important. Next, based on the previous research, the items of multi-functionality of urban agriculture can be classified and organized into three categories, such as economic function (related to employment and industry), environmental function (green space / environmental conservation, landscape formation) and social function (non-economic/ environmental).

Table 5.5 summarized the results of cross tabulation and variable settings for social capital and creative thinking. According to Putnam (1993), social capital is "the characteristics of social institutions such as trust, norms, and networks that improve the efficiency of society by encouraging cooperative action." Although there are systematic researches depending on the purpose and target of the analysis in Japan such as Inaba (2008) and Inaba et al. (2011), there is no established method for the measurement of social capital. In this study, we measure social capital using two questions, "norms of reciprocity" and "general trust", with reference to the previous studies.

First, according to the results of cross tabulation and the chi-square test, there are positive relations among "norms of reciprocity" and "general trust" and "creative thinking" both in Tokyo and Shanghai. And the positive relevance is more clearly shown in Tokyo. However, the index of social capital here is cognitive social capital in the global sense as organized by Uebuchi et al. (2008), and is not limited to the area of residence. The purpose of this study is not to compare the level of social capital between the two regions of Tokyo and Shanghai, but to compare the level of social capital between the people with high and low creative thinking in each region. Therefore, we set the variables of social capital at the individual cognitive level based on these two questions: "Would you say that most of time, people try to be helpful, or that they are mostly just looking out for themselves?" and "Generally speaking, would you say that most people can be trusted or that you had better to be careful in dealing with people?"

[2] Please see the FAO website (http://www.fao.org/urban-agriculture/en/) for more details.

Table 5.4 Cross tabulation about MFA and thinking creatively

			Tokyo							Shanghai						Variables of MFA
			Thinking creatively					Test		Thinking creatively					Test	
			4	3	2	1	0			4	3	2	1	0		
1	Supply of fresh and safe agricultural and livestock products	51.3	58.9	51.2	42.2	54.7	51.5	**	47.8	54.5	43.3	45.1	39.3	50.0	**	SOC
2	Providing the employment	30.4	37.0	31.7	26.7	26.3	30.7		31.7	35.8	29.7	27.1	32.1	0.0		ECO
3	Conservation of the green and environment	46.2	47.4	50.0	45.7	39.5	47.2		56.9	59.0	56.7	53.5	46.4	50.0		ENV
4	Education functions such as farming experience and food education	23.1	23.4	27.0	21.1	21.1	22.1		32.0	35.0	30.8	29.2	25.0	0.0		SOC
5	Revitalization of regional industry	28.0	29.7	34.9	22.8	27.9	22.7	**	31.8	35.0	29.9	29.2	32.1	0.0		ECO
6	Disaster prevention function such as evacuation place in case of disaster	22.2	18.8	26.2	23.7	21.1	9.0		24.7	25.3	23.5	27.1	25.0	0.0		SOC
7	Making a life tasteful and peaceful	30.8	34.9	32.9	30.2	26.3	28.8		38.6	41.0	38.9	31.3	39.3	0.0		SOC
8	Succession of local tradition and culture	19.1	21.9	21.4	16.4	20.5	14.7		29.4	34.5	27.0	21.5	39.3	0.0	**	SOC
9	Formation of good scenery	21.1	26.6	22.2	20.7	19.5	15.3		36.3	36.5	36.7	36.1	32.1	0.0		ENV
10	Place for local community	19.3	23.4	22.6	13.8	17.9	19.0	*	21.1	23.3	18.2	19.4	46.4	0.0	***	SOC

(continued)

Table 5.4 (continued)

			Tokyo							Shanghai						Variables of MFA
			Thinking creatively					Test		Thinking creatively					Test	
			4	3	2	1	0			4	3	2	1	0		
11	Medical and welfare functions such as horticulture therapy	10.7	15.1	13.1	8.2	6.8	9.8	**	21.0	24.5	18.9	16.7	28.6	0.0		SOC
12	Place of familiar recreation	10.8	15.6	10.3	9.5	8.9	9.8		12.1	14.0	9.7	12.5	25.0	0.0	*	SOC
13	Other	0.1	0.0	0.0	0.0	0.0	0.6		0.0	0.0	0.0	0.0	0.0	0.0		–
14	No function seemed important	11.3	7.8	8.3	11.2	15.3	15.3	***	0.4	0.3	0.4	0.0	0.0	50.0	***	–

Note 1 Numerical values by thinking creatively are given as extremely important (4), very important (3), important (2), slightly important (1), and not important (0). Corresponds to things

Note 2 The chi-square test was used for the test. For each item 1–14 (excluding 13), we set a null hypothesis that the response ratio with selection and the response ratio without selection for each importance of creative thinking are the same, and tested. "***", "**", and "*" indicate that there is a significant difference at the 1%, 5%, and 10% levels, respectively (the null hypothesis is rejected)

Note 3 Variables related to the multi-functionality of agriculture (MFA) are numerical values of 1 for each of the economic (ECO), environmental (ENV), and social (SOC) variables when each item from 1 to 12 is selected. The simple accumulation value of each selected item is used as a variable

Table 5.5 Cross tabulation about social capital and thinking creatively

			Tokyo							Shanghai					
			Thinking creatively					Test		Thinking creatively					Test
			4	3	2	1	0			4	3	2	1	0	
Norms of reciprocity	Try to be helpful(1)	59.3	72.4	69.8	60.3	48.9	38.0	***	80.8	84.8	80.2	74.3	67.9	50.0	**
	Look out for themselves(0)	40.7	27.6	30.2	39.7	51.1	62.0		19.2	15.3	19.8	25.7	32.1	50.0	
General trust	Can be trusted(1)	44.6	55.7	47.6	44.8	41.6	30.1	***	77.6	79.3	79.6	69.4	64.3	50.0	**
	Had better to be careful (0)	55.4	44.3	52.4	55.2	58.4	69.9		22.4	20.8	20.4	30.6	35.7	50.0	
Average score(0–2)		1.04	0.64	0.59	0.53	0.45	0.34		1.58	0.82	0.80	0.72	0.66	0.50	

Note 1: The average score is calculated after adding the values in parentheses of the norm of reciprocity and general trust, and simply accumulating them
Note 2 The chi-square test was used for the test. "***", "**", and "*" indicate that there is a significant difference at the 1%, 5%, and 10% levels, respectively (the null hypothesis is rejected)

Table 5.6 summarized the results of cross tabulation and variable settings for creative thinking and their involvement intentions toward social business. In addition, there are various definitions of social business, and conditions as an ideal type have been proposed (for example, Defourny and Nyssens 2006; Matsunaga 2012), but are not fixed. In order to broadly understand SB from various perspectives including entrepreneurs, employees, and users, social business here is defined as "an effort to solve social problems with business methods. Community businesses (CB) that solve local problems with business methods are included". Due to social business has an aspect of creating shared value (CSV), and new value is created by the involvement of various actors (Porter and Kramer 2011), the method of involvement is set from multiple dimensions: entrepreneurship (entrepreneur, investor), staff (paid, free of charge), and users/donors. Specifically, the response items are set with reference to related surveys (Social Business Study Group, 2008; Japan Finance Corporation Research Institute 2014).

According to the results of cross tabulation, there are positive relations between the importance of creative thinking and the presence or absence of intention to be involved (overall) in SB. In Tokyo, there are positive relations between creative thinking and almost all items of intention to be involved (excluding users and others). On the other hand, in Shanghai, there are positive relations between creative thinking and entrepreneurs, investors, and paid staff at the time of establishment, but there are negative relations between free volunteers and users. Based on the above, variables are set as follows: a numerical value (1) is given when there is an intention to be involved, and a numerical value (0) is given when there is no intention to be involved.

Table 5.7 summarized the basic statistics of the variables used in the structural equation modeling. The values in parentheses represent each hierarchy. In the estimation, the correlation of the error terms of each variable in the layer 1, the layer 2, and the layer 3 is assumed. In addition, the maximum likelihood estimation method that takes into account the missing variables is used because there are missing variables for household income among the socio-economic attributes.

5.3 Analytical Results

5.3.1 Analytical Results of Tokyo

Figure 5.2 showed the estimation results for Tokyo (see Table 5.8 for details). The indicators of goodness of fit were Chi-square (3) = 565.894 (p-value = 0.000), RMSEA = 0.427, and CFI = 0.487, respectively. The results of the analysis showed that creative thinking, social capital, multi-functionality of agriculture, and awareness of involvement in social business interact with each other. In the following, we will focus on the items that were significant at the 1% level.

First, in terms of socio-economic attributes and farmland endowment, it was male (+), income (+), and farmland endowment (+) that influenced "creative thinking."

Table 5.6 Cross tabulation about involvement intention of social business and thinking creatively

		Tokyo							Shanghai					
		Thinking creatively					Test		Thinking creatively					Test
		4	3	2	1	0			4	3	2	1	0	
Intention to be involved (1)	46.0	63.0	50.8	39.7	38.4	23.9	***	85.3	90.8	84.2	76.4	75.0	50.0	***
Founder	7.1	19.8	7.9	2.6	3.7	1.2		17.1	24.3	13.6	10.4	7.1	0.0	
Investor	6.1	16.1	5.6	4.3	2.6	1.8		18.0	22.8	16.3	11.1	14.3	0.0	
Paid staff	11.1	17.7	12.7	8.6	9.5	6.1		32.4	32.3	32.1	32.6	39.3	0.0	
Volunteer	8.4	10.9	10.3	9.1	5.3	4.9		25.9	24.5	24.6	31.3	39.3	0.0	
User	21.4	22.9	24.2	21.1	22.6	14.1		17.7	11.8	21.3	19.4	32.1	50.0	
Donor	6.3	8.9	7.9	6.0	4.7	3.1		8.2	9.5	8.1	4.9	7.1	0.0	
Others	0.5	0.5	0.4	0.4	0.0	1.2		0.0	0.0	0.0	0.0	0.0	0.0	
No intention to be involved (0)	56.0	37.0	49.2	60.3	61.6	76.1		14.7	9.3	15.8	23.6	25.0	50.0	

Note The chi-square test was used for the test. We set the null hypothesis that the response ratio of the importance of thinking creatively is the same between "intention to be involved" and "No intention to be involved". "***", "**", and "*" indicate that there is a significant difference at the 1%, 5%, and 10% levels, respectively (the null hypothesis is rejected)

Table 5.7 Descriptive statistics of variables

Variables		Layer	Tokyo					Shanghai				
			Sample	AVG	S.D	Min	Max	Sample	AVG	S.D	Min	Max
Socio-economic attribute	Sex	(1)	1,029	0.76	0.43	0.00	1.00	1,029	0.51	0.50	0.00	1.00
	Age	(1)	1,029	3.64	1.10	1.00	6.00	1,029	2.35	1.10	0.00	6.00
	Income	(1)	879	3.71	2.07	0.00	8.00	1,028	5.67	1.96	0.00	8.00
Farmland		(1)	1,029	1.12	0.93	0.00	3.00	1,029	1.53	0.85	0.00	3.00
Thinking creatively		(2)	1,029	2.12	1.34	0.00	4.00	1,029	3.19	0.79	0.00	4.00
Social capital		(2)	1,029	1.04	0.80	0.00	2.00	1,029	1.58	0.68	0.00	2.00
MFA	ECO	(3)	1,029	0.58	0.69	0.00	2.00	1,029	0.63	0.68	0.00	2.00
	ENV	(3)	1,029	0.67	0.74	0.00	2.00	1,029	0.93	0.71	0.00	2.00
	SOC	(3)	1,029	1.87	1.82	0.00	8.00	1,029	2.27	1.51	0.00	8.00

Note For each variable, the numerical values shown in parentheses below are added to each answer
Sex: Male (1), Female (0)
Age: 18–19 years old (0), 20–29 years old (1), 30–39 years old (2), 40–49 years old (3), 50–59 years old (4), 60–69 years old (5), 70 years old and over (6)
Household income (Tokyo): Less than 2 million yen (0), 2–4 million yen (1), 4–6 million yen (2), 6–8 million yen (3), 8–10 million yen (4), 10–12 million yen or less (5), 12–15 million yen (6), 15–20 million yen (7), 20 million yen and above (8)
Household income (Shanghai): Less than 10,000 yuan (0), 10,000–11,999 yuan (1), 12,000–13,999 yuan (2), 14,000–15,999 yuan (3), 16,000–17,999 yuan (4), 18,000–19,999 yuan (5), 20,000–29,999 yuan (6), 30,000–39,999 yuan (7), 40,000 yuan and above (8)
Farmland endowment: There is a lot (3), There is a little (2), There is not much (1), There is not any(0)
Thinking creatively: Extremely important (4), Very important (3), Important (2), Somewhat important (1), Not important (0)

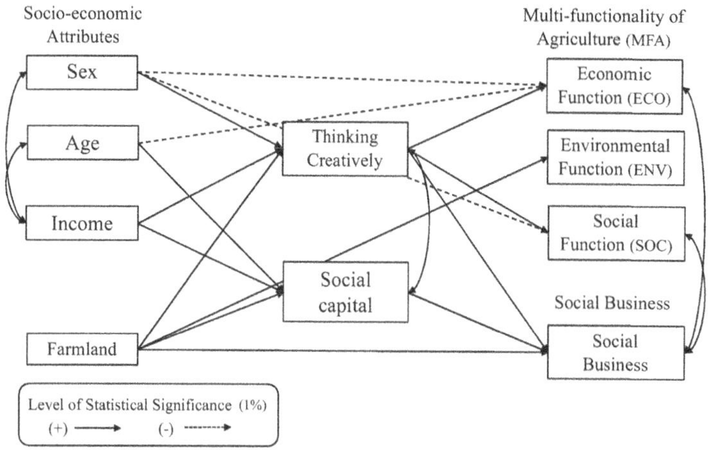

Fig. 5.2 Path diagram (Tokyo)

And, in relation to the multi-functionality of agriculture in the third layer, it is clear that "creative thinking" has a positive influence on economic (+) and social (+) aspects among the multi-functionality of agriculture.

Next, the variables of socio-economic attributes and farmland that affect "social capital" were not confirmed. However, it can be seen that "social capital" has a positive effect on the environmental conservation function (+) on the multi-functionality of agriculture. Furthermore, regarding the relationship between "creative thinking" and "social capital," a positive correlation (+) was confirmed according to the results assuming the covariance of the error terms.

Finally, regarding the relationship with the intention to engage in social business in the third layer, "creative thinking" has a positive influence on the intention to involve social business (+), and social capital also has a positive effect on it. Furthermore, in relation to the multi-functionality of agriculture, a positive correlation was confirmed between economic and social aspects. Therefore, it is clarified that people with creative thinking and high level of social capital have a strong intention to be involved in social business in Tokyo.

5.3.2 Analytical Results of Shanghai

Figure 5.3 showed the estimation results of Shanghai City (see Table 5.8 for details). The indicators of goodness of fit were Chi-square (3) = 62.206 (p-value = 0.000), RMSEA = 0.138, and CFI = 0.862, respectively. Same as the results of Tokyo, the analysis results showed that creative thinking, social capital, multi-functionality of agriculture, and awareness of involvement in social business are influencing each

Table 5.8 Estimated result of SEM (Tokyo and Shanghai)

				Tokyo			Shanghai		
				Coef	P > |z|		Coef	P > |z|	
ECO: MFA	(3)	Thinking creatively	(2)	0.055	0.001	***	0.049	0.068	*
	(3)	Social capital	(2)	0.026	0.354		0.039	0.205	
	(3)	Farmland	(1)	0.053	0.024	**	0.075	0.003	***
	(3)	Sex	(1)	−0.162	0.002	***	−0.106	0.013	**
	(3)	Age	(1)	−0.054	0.007	***	0.004	0.837	
	(3)	Income	(1)	−0.023	0.061	*	0.021	0.051	*
ENV: MFA	(3)	Thinking creatively	(2)	0.044	0.016	**	0.033	0.247	
	(3)	Social capital	(2)	0.006	0.840		0.128	0.000	***
	(3)	Farmland	(1)	0.068	0.007	***	0.006	0.815	
	(3)	Sex	(1)	−0.145	0.011	**	−0.080	0.075	*
	(3)	Age	(1)	0.028	0.187		−0.036	0.091	*
	(3)	Income	(1)	−0.007	0.606		−0.017	0.148	
SOC: MFA	(3)	Thinking creatively	(2)	0.152	0.001	***	0.182	0.003	***
	(3)	Social capital	(2)	0.182	0.013	**	0.034	0.620	
	(3)	Farmland	(1)	0.091	0.132		0.120	0.033	**
	(3)	Sex	(1)	−0.850	0.000	***	−0.310	0.001	***
	(3)	Age	(1)	0.058	0.259		−0.023	0.615	
	(3)	Income	(1)	−0.027	0.399		−0.007	0.790	
Social business	(3)	Thinking creatively	(2)	0.069	0.000	***	0.036	0.004	***
	(3)	Social capital	(2)	0.107	0.000	***	0.044	0.002	***
	(3)	Farmland	(1)	0.069	0.000	***	0.161	0.000	***
	(3)	Sex	(1)	−0.055	0.125		−0.023	0.244	
	(3)	Age	(1)	−0.011	0.412		0.028	0.003	***
	(3)	Income	(1)	0.009	0.268		0.030	0.000	***
Thinking creatively	(2)	Farmland	(1)	0.178	0.000	***	0.095	0.001	***
	(2)	Sex	(1)	0.688	0.000	***	0.032	0.519	
	(2)	Age	(1)	−0.035	0.341		−0.040	0.090	*
	(2)	Income	(1)	0.115	0.000	***	0.063	0.000	***
Social capital	(2)	Farmland	(1)	0.134	0.000	***	0.034	0.190	
	(2)	Sex	(1)	0.138	0.018	**	0.029	0.501	
	(2)	Age	(1)	0.088	0.000	***	−0.008	0.692	
	(2)	Income	(1)	0.059	0.000	***	0.019	0.081	*

(continued)

Table 5.8 (continued)

				Tokyo			Shanghai		
				Coef	P > |z|		Coef	P > |z|	
Covariance of error terms									
Social business	(3)	ECO: MFA	(3)	0.030	0.007	***	0.006	0.370	
	(3)	ENV: MFA	(3)	0.021	0.070	*	0.000	0.982	
	(3)	SOC: MFA	(3)	0.162	0.000	***	0.035	0.019	**
Social capital	(2)	Thinking creatively	(2)	0.178	0.000	***	0.054	0.001	***
Farmland	(1)	Sex	(1)	0.023	0.068	*	−0.005	0.714	
	(1)	Age	(1)	0.036	0.263		0.168	0.000	***
	(1)	Income	(1)	0.044	0.494		−0.027	0.604	
Sex	(1)	Age	(1)	0.026	0.076	*	0.125	0.000	***
	(1)	Income	(1)	0.205	0.000	***	0.014	0.646	
Age	(1)	Income	(1)	0.397	0.000	***	0.421	0.000	***

Note "*", "**" and "***" indicate that they are statistically significant at the 10%, 5% and 1% levels, respectively

MFA stands for Multi-functionality of Agriculture

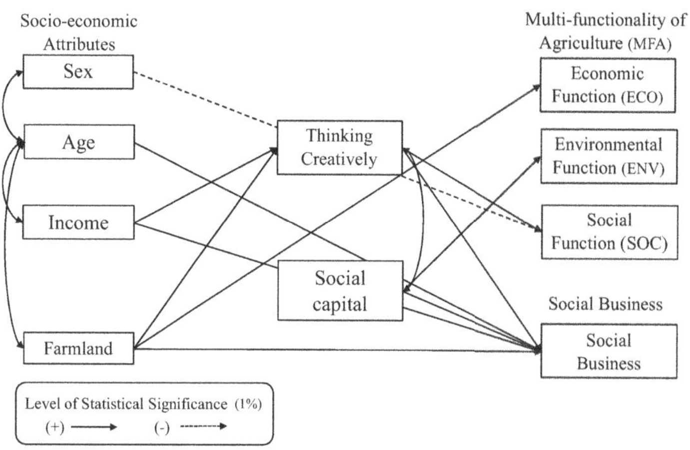

Fig. 5.3 Path diagram (Shanghai)

other. In the following, we will focus on the items that were significant at the 1% level.

First, in terms of socio-economic attributes and farmland endowment, it was income (+) and farmland endowment (+) that influenced "creative thinking." And, in relation to the multi-functionality of agriculture in the third layer, "creative thinking" has a positive influence on sociality (+).

Next, the variables of socio-economic attributes and farmland that affect "social capital" were not confirmed. However, it can be seen that "social capital" has a positive effect on the environmental conservation function (+) on the multi-functionality of agriculture. Furthermore, regarding the relationship between "creative thinking" and "social capital," a positive correlation (+) was confirmed according to the results assuming the covariance of the error terms.

Finally, it was seen that "creative thinking" and social capital had a positive impact on the involvement in social business (+). However, no positive correlation was found for SB involvement in relation to the multi-functionality of agriculture. Therefore, in Shanghai, it is clarified that people with creative thinking and high social capital level have a strong intention to be involved in social business. However, it differed from the results of Tokyo in that the relationship between the multi-functionality of agriculture and social business was not confirmed.

5.4 Conclusion and Policy Implication

Based on the above analytical results, we obtained the following conclusions regarding the relations between multi-functionality of agriculture and the creative classes in Tokyo and Shanghai. In short, the people who are thinking creatively in the two cities prefer the multi-functionality of agriculture more than other people, the trend is more pronounced in Tokyo. And those who value creative thinking have a high level of social capital and are more willing to get involved in social business. However, in Shanghai, we could not find a connection between the multi-functionality of agriculture and social business like in Tokyo. This is thought that social business has not yet developed in China due to the immaturity of civil society (Li and Merkle 2017).

In today's fierce competition between global cities, there are various measures for attracting creative classes. However, they are generally the idea of zero-sum games, and many of them are transient with side effects. Brain competition policy (BCP) is based on the complexity of factors that shape the mobility and location decisions of highly-skilled workers. And it is about the attraction, education, circulation and utilization of talent functioning as knowledge spillover agents in and between regional, national and supranational economies (Reiner 2010). On the other hand, solving social issues through social business can be considered as social innovation in aiming for an outstanding global city. The attempts to solve food security, environmental and disparity/isolation issues through urban agriculture are seen in practice in the two cities of Tokyo and Shanghai.

Therefore, by involving creative class in urban agriculture as a social business, some social problems such as lack of fresh and safe agricultural products, weakening of local communities, deterioration of natural environment, lack of recreation, lack of measures against natural disasters and decline of local traditions etc. in megacities can be solved to a certain extent.

References

Defourny J, Nyssens M (2006) Defining social enterprise. In: Nyssens M (ed) Social enterprise: at the crossroads of market, public policies and civil society. Routledge, pp 3–26

Inaba Y (ed) (2008) Potentials of social capital. Nippon Hyoron Sha, Tokyo (in Japanese)

Inaba Y, Ohmori T, Kondo T, Miyata K, Yano S, Yoshino R (2011) Frontiers of social capital: achievements and possibilities. Minerva Shobo, Kyoto (in Japanese)

Japan Finance Corporation Research Institute (2014) Questionnaire survey on the social business and community business. (http://www.jfc.go.jp/n/findings/pdf/sme_findings140912.pdf) (in Japanese)

Kiminami L, Kiminami A (2006) Sustainability of agriculture and urban quality of life in Japan—economic efficiency, sociality and environment protection. Stud Region Sci 36(2):305–321

Kiminami L, Kiminami A (2007) Sustainability of urban agriculture: a comparative analysis of Tokyo and Shanghai. Stud Region Sci 37(2):585–597

Kiminami L, Kiminami A, Furuzawa S (2018) Impacts of multi-functionality of urban agriculture on the CCs in Japan. Asia-Pacific J Region Sci 2(2):507–527

Kiminami L, Furuzawa S, Kiminami A (2020) Empirical study on the multi-functionality of urban agriculture and creative classes: focusing on the global cities of Tokyo and Shanghai. Stud Region Sci 50(1):73–96 (in Japanese)

Li D, Merkle A (2017) Talent: the key to developing the social sector in China. Stanford Social Innov Rev. https://ssir.org/articles/entry/talent_the_key_to_developing_the_social_sector_in_china

McGranahan D, Wojan T (2007) Recasting the creative class to examine growth processes in rural and urban counties. Reg Stud 41(2):197–216

Matsunaga Y (2012) Quantitative analysis on social enterprises. In: Quantitative analysis on new public and civil society, Osaka University Press, Osaka, pp 113–141 (in Japanese)

OECD (2001) Multifunctionality: towards an analytical framework. OECD Publishing, Paris

Porter ME, Kramer MR (2011) The big idea: creating shared value (How to reinvent capitalism—and unleash a wave of innovation and growth). Harv Bus Rev 89(1–2):62–77

Putnam R (1993) Making democracy work: civic traditions in modern Italy. Princeton University Press, Princeton

Reiner C (2010) Brain competition policy as a new paradigm of regional policy: a European perspective. Pap Reg Sci 89(2):449–461

Social Business Study Group (2008) Report of Social Business Study Group. https://www.meti.go.jp/policy/local_economy/sbcb/sbkenkyukai/sbkenkyukaihoukokusho.pdf (in Japanese)

Uebuchi T, Ichida Y, Hirai H, Kondo K (2008) Social capital and region: issues on the empirical analysis at the regional level. In: Inaba Y (ed) Potentials of social capital. Nippon Hyoron Sha, Tokyo, pp 55–72 (in Japanese)

Chapter 6
Social Entrepreneurship and the Multi-functionality of Urban Agriculture in Japan and China

Abstract The purpose of this chapter is to clarify the mechanism of social entrepreneurship in urban agriculture, and examine the relationships among social institutions, social business and the multi-functionality of urban agriculture in Japan and China. The method of trajectory equifinality modeling (TEM) is introduced to each three cases from Tokyo (Japan) and Shanghai (China). The results indicated that social entrepreneurship (SE) was affected by social institutions (SI) on one side, and it also enhanced changes in social institution through the activities of social business (SB) on the other. And social capital (SC) was critically important to social entrepreneurship for the development of social business. Furthermore, it is expected that the residents will have a deeper understanding and a more active involvement in urban agriculture with the development of social businesses that bring out the multi-functionality of agriculture if more stakeholders are involved.

Keywords Trajectory equifinality modeling (TEM) · Social institution (SI) · Social entrepreneurship (SE) · Social capital (SC) · Social business (SB) · Tokyo · Shanghai

6.1 Introduction

The analytical framework of this chapter is shown by Fig. 6.1. Social institutions affect social entrepreneurship (SE) on one side and receive influences from social business (SB) on the other. In other words, one of the characteristics of social entrepreneurship is that it has a path (A → B → C) to change social institutions through social business. Here, social institutions (SI) involve relationships among individual members of society. Economic transactions themselves are social relationships, but not all social relationships are economic transactions. The members of society form social institutions either deliberately through explicitly cooperative agreements and implicitly competition, or even informally through "spontaneous order". Institutions include political mechanisms and other forms of social order, as well as evolving social conventions such as language, rules, customs, kinship, community traditions, cultural values, and behavioral norms (Spulber 2014).

L. Kiminami et al., *Social Entrepreneurship, Social Business and the Multi-functionality of Urban Agriculture*, SpringerBriefs in Economics,
https://doi.org/10.1007/978-981-19-1762-2_6

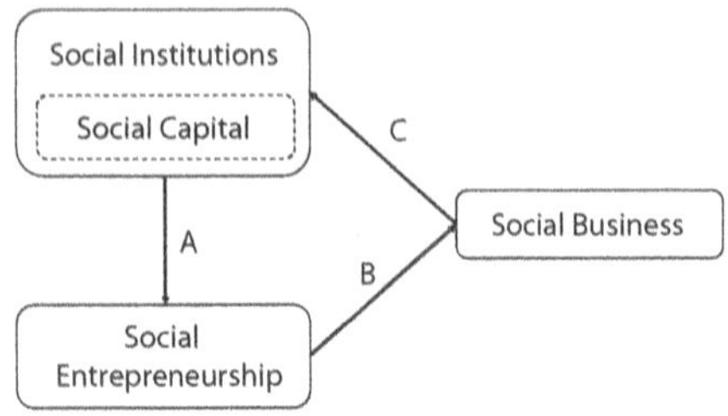

Fig. 6.1 Analytical framework

In the following, we will clarify the development mechanism of social entrepreneurship in urban agriculture in Tokyo and Shanghai by using the TEM analysis. Table 6.1 summarized the overview of 6 cases of social entrepreneurs (3 cases for Tokyo and 3 cases for Shanghai). Materials used for these case analyses are shown in Appendix 6.1.

6.2 TEM Analysis for Tokyo

6.2.1 Results of Mr. O's TEM Analysis

Located in Kunitachi City of Tokyo, a community farm (Farm K) operated by Mr. O has its size of 1000 m^2 field. However, there are a small paddy field, vegetable fields, a BBQ space, earthen pipes, handmade playground equipment and animals (ponies) in the farm. The number of users is more than 5000 per year. The diagram of TEM for the case of Mr. O is shown by Fig. 6.2 and the explanations on BFP, SIG and SID are as shown in Table 6.2.

Mr. O was born in Kanagawa Prefecture. After graduating from university, he got a job at a TV station and was involved in program planning and production. The environmental destruction of the forests in the Philippines he visited during the program filming raised his interest in environmental issues (SIG1). Since then, he started to pay strong attention to the producing of agricultural products in his daily life, and determined to get a job involved in agriculture.

As the first step, Mr. O changed his job to an agricultural corporation (BFP1) for learning techniques of farming and knowing about the situation of Japanese agriculture. Eventually, he realized that agriculture in urban area has important social significance (SIG2) where many people know little about production site of agriculture. After that, he moved to another agricultural corporation (BFP2). While working as an employee, he started a private rental "BBQ Farm" in 2010. Although the farm was focusing on the space design of multi-purposes, the concept was unclear and the property tax for housing land was imposed, so the nature as a social business was weak and the sustainability as a business was also low.

In 2013, Mr. O established a community farm "Kunitachi Hatakenbo" (BFP3) with support from Kunitachi City (SIG3). The concept of the farm was clarified through

Table 6.1 Overview of analyzed cases in Tokyo and Shanghai

	Mr. O	Mr. M	Mr. N	Mr. L	Mr. Z	Mr. W
Location	Tokyo	Fujisawa City (Tokyo Metropolitan area)	Tokyo and other big cities	Shanghai	Shanghai	Shanghai
Business	Community garden	Livestock farming	Agricultural experience, etc	Community gardens	Organic Farming	Rooftop Farm
Business type	NPO	Joint-stock company	Joint-stock company	NGO	Limited company	Limited company
Way of entry	New entry (Outside the area)	Business Succession	New entry (Outside the area)	New entry	New entry	New entry
Development of social business	High	High	High	High (Start-up stage)	High (Diversification stage)	High (Expansion stage)
Social value	Service for experiencing agriculture and food (education, community interaction)	Human cultivation (Social entrepreneurs) Set up research group for family business	Agricultural experience service, farming support, elimination of abandoned cultivated land	Improving the community environment and strengtheningties between residents	Providing safe organic vegetables to urban residents, promoting the concept of organic and agriculture, and developing various projects such as agritourism	The business makes effective use of the rooftop space of a commercial building for residents to interact and engage in food education activities
Changes in social entrepreneurship	Rising	Rising	Rising	Rising	Rising	Rising

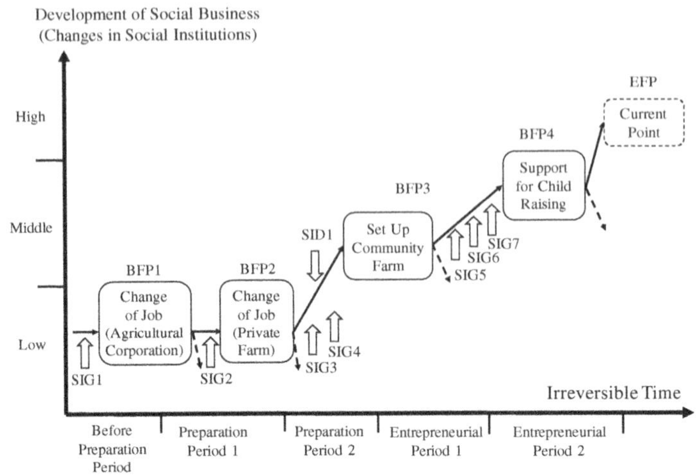

Fig. 6.2 TEM diagram of the analysis on Mr. O

Table 6.2 List of Mr. O's BFP, SIG and SID

	Contents
BFP1	Changing his job to an agricultural corporation
2	Moving to another agricultural corporation
3	Establishing a community farm "Kunitachi Hatakenbo"
4	Starting child-rearing support project
SIG1	Visiting the Philipines during the program filming raised his interest in environmental issues
2	Realizing social significance of urban agriculture
3	Support from Kunitachi City
4	Determining the farm's concept through repeated consultations with the government and local stakeholders
5	Cooperation with MAFF, Kunitachi City
6	Joint activities with local citizens
7	Amendment of "Production Green Space Law"
SID1	The previous "Production Green Space Law"

repeated consultations with the government and local stakeholders (SIG4). The farm is an entertainment-type agricultural service for providing agricultural experience that includes elements for parents and children, spouse hunting, and Ninja experience. The philosophy of the farm has the following three points: (1) Diversified use (both publicly and profitably); (2) Openness of integrating the surrounding environment and local communities; and (3) Offering a place to citizens for farmland using. In other words, it is an innovative social business that provides space and services with

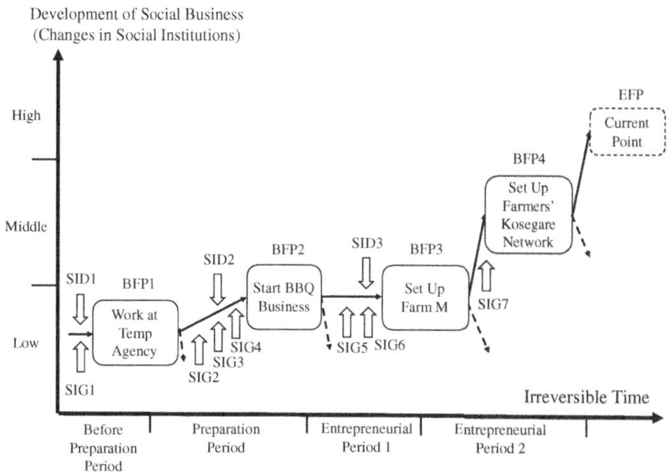

Fig. 6.3 TEM diagram of the analysis on Mr. M

diversity, openness and comfort to communities. However, the diversification of farm business was prohibited by the previous "Production Green Space Law" (SID1).

In 2017, a local child-rearing support project (BFP4) was started using old farmhouse in cooperation with MAFF, Kunitachi City (SIG5) and joint activities with local citizens (SIG6) in response to the amendment of "Production Green Space Law'[1] (SIG7). Mr. O is now working to expand his business (EFP) as Tokyo Urban Agritourism by accepting private overnight stays and inbound tourists using old farmhouses.

6.2.2 *Results of Mr. M's TEM Analysis*

Farm M is a pork farming company located in Fujisawa City, Kanagawa Prefecture. In addition to the pork production business, Farm M conducts diversification of agri-business such as processing, direct sale and BBQ business (interacting between producers and consumers). The diagram of TEM for the case of Mr. M is shown by Fig. 6.3 and the explanations on BFP, SIG and SID are as shown in Table 6.3.

Mr. M had been thinking about starting a business in the future since he was a university student (SIG1), but he was not interested in agriculture, and was not willing to succeed the pork farming run by his father (SID1). After graduated from university, he found a job at a temporary staffing company (BFP1). In April 2001, he began studying about agriculture and recognized the problems and appeal of agriculture (SIG2). Aiming to make the primary industry a cool, impressive, and profitable one,

[1] Which has made possible to set up farmer's restaurants and direct sales offices in production green spaces in urban areas.

Table 6.3 List of Mr. M's BFP, SIG and SID

	Contents
BFP1	Found a job at a temporary staffing company
2	Started a BBQ business
3	Established farm M
4	Starting activities to help people participate in their own family agriculture
SIG1	Thinking about starting a business in the future since he was a university student
2	Began studying about agriculture himself and recognized the problems and appeal of agriculture
3	Concentrated on the sales department was established
4	Support of friends and acquaintances from the networks he has built so far
5	Started the transmission of local information
6	Received support for entrepreneur development
7	Collaborating with companies that connect the agricultural industry with university students
SID1	Not willing to succeed the pork farming run by his father
2	Mr. M's proposal was strongly opposed by his father
3	The necessity to promote branding of its own pork production and establish it as a business

he started the preparations for a business. After leaving from the company, he went back to his parents' home and envisaged a business to revitalize the local area through agriculture. To that end, he proposed to his father to set up a sales business for the produced pork. However, his father had a traditional management policy and strongly opposed his proposal (SID2). On the other hand, his younger brother agreed with his proposal and joined the production department of pork farming in April 2005. A system that Mr. M could concentrate on the sales department was established (SIG3).

In June 2005, Mr. M participated in his family's pork farming management and started a BBQ business in August (BFP2). With the support from the network of his friends and acquaintances (SIG4), the BBQ business was smoothly on track. In the process of participating in activities to tackle local problems, his understanding about the actual issues of regional agriculture was deepening. He started the transmission of local information (SIG5) instead of implementing the businesses for butchers, restaurants and processing.

From August 2006 to April 2007, Mr. M's corporate plan was adopted by the NEC Social Entrepreneurship School and received support for entrepreneur development (SIG6). However, the necessity to promote branding of its own pork production and establish it as a business was pointed out (SID3) as a social entrepreneur. A direct sales business was started in cooperation with a meat-processing company. In September 2006, Farm M was established (BFP3). The company promoted the branding of pork by selling pork to restaurants and selling to individuals by ordering on the web.

Based on his own experience, Mr. M began to realize that the challenge of Japanese agriculture is business succession. By collaborating with companies and universities that connected to the agricultural industry (SIG7), he started to demonstrate the attractiveness of agriculture and help the young people from farm households to participate in their own family agriculture (BFP4). In 2015, he built a family business study group to provide a platform to smoothly convey business succession methods and the appeal of family business (EFP).

6.2.3 Results of Mr. N's TEM Analysis

MYFARM Inc., is the main example of start-ups of Mr. N founded in 2007 with the idea of "self-production for local consumption" (producing and eating by ourselves). The main business is an agricultural experience service, which is operated by utilizing abandoned cultivated land all over the country (reaching 100 locations in 2015). In addition, it has developed various businesses such as an agricultural school (Agri-Innovation College) that is open on weekends, an online wholesale market, an agricultural management consulting business, a mail order business, and research and development. The diagram of TEM for the case of Mr. N is shown by Fig. 6.4 and the explanations on BFP, SIG and SID are as shown in Table 6.4.

Mr. N enjoyed vegetable cultivation since the childhood. At high school, he was interested in the problem of cultivated abandoned farmlands on the way to school (SIG1). From university third grader, he studied a variety improvement and knew that it takes time to commercialize because it is necessary to have a variety registration and law revision (SID1). While examining farmland law etc., he thought about

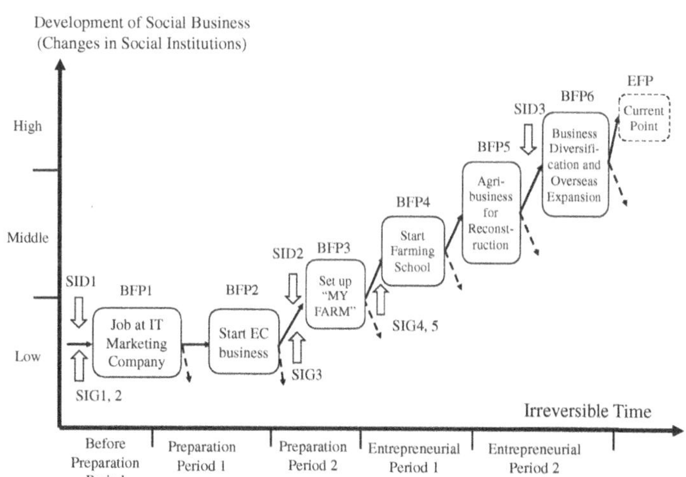

Fig. 6.4 TEM diagram of the analysis on Mr. N

Table 6.4 List of Mr. N's BFP, SIG and SID

	Contents
BFP1	Got a job at an IT marketing company
2	Started a sales business of craft products by internet (EC)
3	Started MYFARM (agricultural experience farm)
4	Starting farming support business in my farm academy (2010 onward)
5	Launched an agricultural business related to Tohoku reconstruction
6	Developed various agriculture-related businesses, and expanded overseas (Shanghai, China)
SIG1	Growing vegetables in the childhood, and had an interest in issues of abandoned farmland on the way to school
2	Thought about starting up agricultural experience business while examining farmland law etc
3	Encountered entrepreneurial partner Mr. I at an incubator office
4	Found a cooperating farmer through acquaintances and friends, and met local young farmers
5	Enlarged the business, and entry of imitative businesses and regional revitalization
SID1	Knew that the commercialization of variety improvement takes time because of a regulation and law
2	Looked around the farmers to cooperate (visiting 300 locations in half a year) but no owner who lends unused farmland
3	The harvest of trial cultivated tomato was successful and called topics, but it received criticism from the scientific basis

starting up agricultural experience business (SIG2). Also, as part of preparation for entrepreneurship, he found a job in an IT marketing company (BFP1).

He resigned the job and worked as an intern in IT venture company in Kyoto for a month, and launched the company "OKOSHI" for selling the traditional crafts by internet (BFP2). At that intern destination, he met Mr. I who was a co-founder of My FARM INC. and the current director of it (SIG3). Mr. I also felt the possibility of social entrepreneurship and they discussed about the mission of MY FARM INC. At that point, "Increase the number of people who can make their own food" and "Regenerate abandoned cultivated land". Three stages (1: Experience farms, 2: Academy business, 3: Field creation business) were set to develop a concrete plan for our business.

For starting agricultural experience businesses, he searched nearly 300 farmers for cooperation in half a year, but no owner would like to lend unused farmland (SID2) to him. Without giving up, he finally got to know the young representative of Japan Agriculture Association (who has a pipe with the Ministry of Agriculture) and local young farmers (SIG4) at the introduction of acquaintance.

In April 2008, he started 50 compartments (1 compartment 15 m^2) for agricultural experience in Kyoto (BFP3), and encountered organic agricultural experts and asked for cultivation guidance. As a result of a lot of similar businesses, a large farm area was formed near the farm, and a lively community was born (SIG 5). In 2010, a farm academy starts in Yasu City to support farming (BFP 4). In order to cooperate with the "problem of salt damage" caused by the tsunami in 2011, he launched "Reconstruction tomato project" (BFP 5) collaborated with local governments, NPOs and universities. In August 2011, the harvest of trial cultivated tomato was successful and called topics, but it received criticism from the scientific basis (SID3). In 2013, the business of "Distribution Innovation" to deliver fresh and inexpensive agricultural products to consumers was started (BFP 6).

In recent years, various businesses have been developed by MYFARM Inc. (started with a beekeeping in 2014, a comprehensive collaboration agreement with Tokyo Agricultural University in 2015; service for household vegetable users using social media, school business for farm management, and text book publishing of Agri-Innovation College, etc. in 2016). Furthermore, in 2018, overseas development was also conducted in Shanghai (MYFARM CHINA INC.).

6.3 TEM Analysis for Shanghai[2]

6.3.1 Results of Mr. L's TEM Analysis

Mr. L has been engaged in teaching and research on Sustainable Landscape Planning and design for many years in Shanghai. In recent years, based on urban communities, he has promoted and explored residents' voluntary community landscape construction and management through nature education and community gardens (CGs). Although it is confirmed that the need for professional guidance in the early stage of CGs, Mr. L's business has always emphasized the autonomy of local residents to participate in the construction and operation by mobilizing, guiding and motivating. The business is processed into several stages: testing CGs' projects in practice and then gradually spreading them throughout China.

The diagram of TEM for the case of Mr. L is shown by Fig. 6.5 and the explanations on BFP, SIG and SID are as shown in Table 6.5. The irreversible time was divided into four stages such as "Childhood/School days", "Preparation period 1", "Preparation period 2" and "Entrepreneurial period" of the CGs project.

Mr. L came from a village in Qingdao City of Shandong Province, China. When he was a child, he had a lot of experience in agricultural work which was very hard for him (SID1). Therefore, he went to the Architecture and Urban Planning of Shanghai T University to study Landscape Architecture (BFP1).

When Mr. L worked as a designer for five years after university graduation (BFP2), he saw many huge commercial landscapes (e.g. large fountains) built with a lot of money, but abandoned only a few years later (SID2). In addition, while trying to adapt to the life of Shanghai, Mr. L started to be aware of the plight of urban residents,

[2] Original version of the contents in this section was presented at 58th Annual Conference of JSRSAI: Ge Y, Kiminami L, Furuzawa S "An analysis of social entrepreneurship and innovative agri-business for sustainable urban development: Case studies from Shanghai, China" (in Japanese).

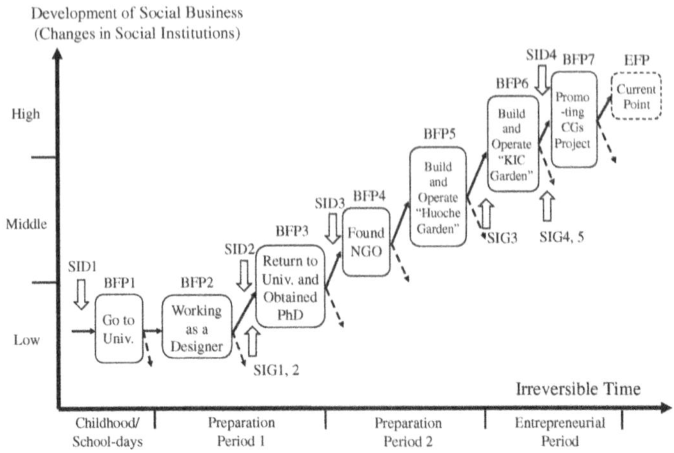

Fig. 6.5 TEM diagram of the analysis on Mr. L

Table 6.5 List of Mr. L's BFP, SIG and SID

	Contents
BFP1	Entering T university in Shanghai
2	Working as a designer after graduation
3	Returning to T university and obtaining PhD
4	Establishing the NGO "Clover Nature School"
5	Building and operating "Huoche Garden"
6	Building and operating "KIC Garden"
7	Promoting CGs projects
SIG1	Starting to be aware of the plight of urban residents
2	Recognizing the needs of urban residents for agriculture
3	A sustainable urban development mode with the emphasis of small-scale renovation
4	Policy for waste separation and recycling in Shanghai
5	Participation of local residents and non-local people in various activities
SID1	Experience in hard agricultural work when he was a child
2	Waste of resources due to unreasonable urban landscape
3	Proposal to build an urban farm was not accepted
4	Inactive participation of residents other than the elderly and children

such as serious air pollution, poor natural space, weak interpersonal relationships and few opportunities for children access to nature, and that the solution to these problems lies in the experience of his rural life (SIG1). On the other hand, he realized that the needs of urban residents for agriculture were not met. Some residents voluntarily cultivated the community's public spaces with crops, but they were prone to conflict due to the lack of reasonable operation (SIG2). He returned to T University and obtained his PhD in urban planning and design (BFP3). In the next decade, Mr. L proposed to operate and maintain urban farms and integrate them into various landscape plans, however his proposal was never accepted by developers because "urban farms are not beautiful" and "construction time is too long" (SID3). In early 2014, Mr. L decided to build community gardens (CGs) together with his friends without giving up, and set up "Clover Nature School" (BFP4). In the same year, they built and operated the "Huoche Garden", an urban farm project (BFP5). At the same time, the Shanghai municipal government realized that the old method of urban development (large-scale demolition and reconstruction) is not sustainable, and is looking for a sustainable urban development mode, with increasing emphasis on small-scale renovation of old towns (SIG3). Under such circumstance, Mr. L led the team to comprehensively promote CGs projects (such as "Knowledge and Innovation Community Garden") in Shanghai (BFP6), with the full support of different stakeholders such as the Shanghai Municipal Government, the residents' committees and socially responsible enterprises.

Mr. L and his team pay special attention to the principle of sustainability when they build CG. The facilities provided by CGs include garbage sorting bin, earthworm tower, various composting facilities, rainwater collection bucket and a small greenhouse, etc. With the official implementation of the *Regulations of Shanghai Municipality on Municipal Solid Waste Management* on July 1, 2019 (SIG4), communities and families in Shanghai are requested to follow the classification of residual waste and household food waste. A large amount of kitchen waste has more opportunities to be used scientifically.

Due to the lack of understanding and construction capacity, there are few active participating residents at the initial stage of construction of CGs except for the elderly and children (SID4). However, by improving operation under professional guidance and organizing various activities, CGs have attracted more residents including the people outside the local area (SIG5) for participation. Finally, under the COVID-19 epidemic, Mr. L and his team are attempting to rebuild the connection and trust between people and between people and communities by advocating non-contact ways to share seeds and plants (BFP7).

6.3.2 Results of Mr. Z's TEM Analysis

Mr. Z's Tony's Farm is relatively large in scale and epitomizes the construction of a distinctive organic food base in Shanghai. It played a role in providing organic vegetables to a wide range of consumers through diversified operation, forming a model business complex with agricultural, leisure and agritourism functions. Mr. Z's organic agriculture business requires a high degree of expertise in soil, water and farming, a good relationship with the government and the formulation of appropriate business strategies. In terms of operation, Mr. Z realized the cooperative relationship between the company and universities and the government, attracted human resources outside the agricultural field to participate in the company's management, developed Tony's Farm into one of the leading organic farming enterprises in China.

The diagram of TEM for the case of Mr. Z is shown in Fig. 6.6 and the explanations on BFP, SIG and SID are as shown in Table 6.6. The irreversible time was divided into four stages, such as "School days", "Preparation period1", "Preparation period 2" and "Entrepreneurial period" of the organic farming business. Considering the limitations of TEM analysis, the time horizon of this analysis will be up to the middle of Mr. Z's start-up of his organic farming business. The reason is that when the scale of the organization is getting large, such as in a group enterprise, the key decisions of the enterprise become more complex, and it is difficult to apply TEM analysis mainly involving individual decisions.

Mr. Z was born in a farm family, graduated from Sichuan N University (Agronomy) (BFP1), and worked as a township head in a mountainous area near the border of Yunnan and Sichuan for three years. However, Mr. Z resigned from his job in the government and state-owned enterprises and engaged in international trade individually (BFP2). In 1991, he read from newspaper that Shanghai was preparing to develop Pudong New Area (SIG1) and decided to go to Pudong for business chance (BFP3), but it was against by his family (SID1). After he had adapted to the local business environment in Shanghai, he started to miss the delicious food of his hometown. He opened the first "Duoli Sichuan restaurant" in Shanghai in 1997 (BFP4). Mr. Z's catering industry was very successful, with more than 30 branches in Shanghai (SIG2). It not only provided him the funds for starting his own farm later, but also helped him succeed in the organic farming industry. The deterioration of the food in the restaurant he operated made him realize the problems of raw materials and supply chain. Mr. Z became very aware of the high demand for healthy and safe food in society, and an emerging market with great potential (SIG3).

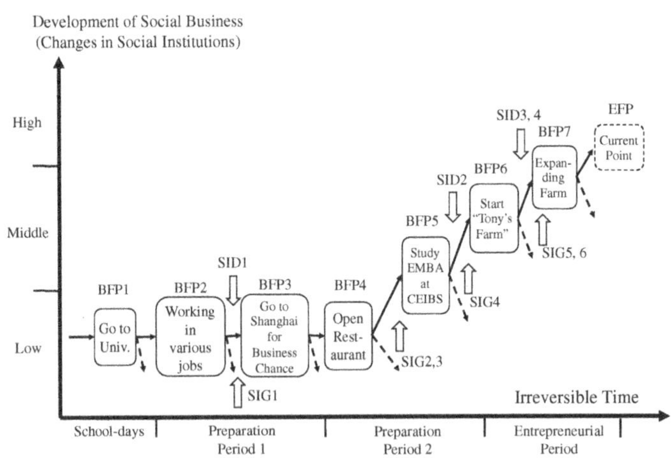

Fig. 6.6 TEM diagram of the analysis on Mr. Z

Table 6.6 List of Mr. Z's BFP, SIG and SID

	Contents
BFP1	Studied at Sichuan N University
2	Worked in various jobs after graduation
3	Went to Pudong (Shanghai) for business chance
4	Opened "Duoli Sichuan restaurant"
5	Studied EMBA at CEIBS
6	Started Tony's Farm
7	Expanding Tony's Farm
SIG1	Shanghai was preparing to vigorously develop the Pudong New Area
2	The success of the catering industry has brought capital
3	Aware of growing demand for healthy and safe food in the market
4	Shanghai launched the plan to establish vegetable production bases
5	New strategies developed by bringing together various talents
6	Venture capitalists are interested in the potential of organic farming
SID1	Opposition from all family members
2	Sell catering business and take significant investment risks in new business
3	Faced with the costs of soil improvement and construction plans
4	Problems with production capacity

In 2005, he chose "urban agriculture (organic farming)" as his major topic to study EMBA (Executive Master of Business Administration) at CEIBS (China Europe International Business School) (BFP5). After completing his EMBA, Mr. Z started his own organic farming business, Tony's Farm (BFP6). To do this, he boldly sold his Sichuan restaurant, which had brought him huge profits (SID2), and invested all his money in improving the environment of the organic farming base, which he did not see any return for many years. Initially, the Shanghai government had just launched a plan to build 5–10 ha vegetable production bases (SIG4). Mr. Z's strong economic strength and his management concept (to produce and sell organic vegetables to the middle class and wealthy people in urban areas, and provide them with the opportunity to experience agriculture on the farm) are consistent with the government's position on land use, which enabled him to obtain the management right of 1,750 acres of farmland in Datuan Town, Pudong New Area. However, the base of organic production needed to carry out soil remediation and construction projects (SID3), which put great pressure on him. Mr. Z actively cooperated with universities and researchers in different regions when facing the problem of production capacity (SID4). He decided to find a way different from the traditional agricultural model, and has convened a variety of talents to formulate management strategies that are unfamiliar in agriculture (SIG5).

At the beginning of Tony's Farm, the channel management principle of "compressing intermediate links" was determined, and the direct sales mode and the member pre-sales mode were finally chosen. In addition, a management pattern was formed at Tony's Farm, with experts providing consultancy, senior managers controlling total production, mid-level employees managing farmers and pest control, and foundation farmers doing the planting. On the other hand, the frequent food safety problems have made organic farming a popular industry sought after by venture capitalists (SIG6).

By introducing membership into their business model, they have succeeded in bringing together urban people with the same philosophy of "organic" meaning healthy and environmental (BFP7), and laid a solid foundation for subsequent significant growth. On the other hand, Mr. Z's organic farming business is not only about production and sales, but also about improving the local environmental infrastructure and providing ecological services and commercial services such as agritourism.

6.3.3 Results of Mr. W's TEM Analysis

Mr. W is the owner of a rooftop farm in Shanghai. He opened a travel agency (Datong travel) and then ran a rooftop farm called "Little bee farm" with his partners. The unique feature of Mr. W's rooftop farm is the effective use of the roof of the building as a space. In addition to improving environmental conditions and beautifying the landscape, rooftop farms can also enable urban residents to feel and experience agriculture more closely than ordinary roof greening. On the other hand, opening a farm in a large shopping mall makes it easier to attract their own customers, and brings publicity to the mall through providing it with new features, ultimately achieving a win–win relationship.

The diagram of TEM for the case of Mr. W is shown in Fig. 6.7 and the explanations on BFP, SIG and SID are as shown in Table 6.7. The irreversible time was divided into three stages, such as "Preparation period", "Preparation period 2" and "Entrepreneurial period" of the rooftop farm business.

Mr. W resigned a stable job (SIG1) and went to Shanghai for business chance (BFP1) with 10,000 yuan in his pocket against his parents' wishes (SID1) when he was 26 years old. He started tourism business in Shanghai for promoting tourism of his hometown Anji (Huzhou City, Zhejiang Province) (BFP2). However, as Anji was not popular as a tourist destination (SID2) at the time, he spent 2 years to obtain the first fund (SIG2) by operating Anji tourist route. In January 2003, he opened his own travel agency, "Datong travel" (BFP3). The company's business radius rapidly spread from East China to the whole country although the rent of office in Huangpu District, Shanghai at that time was more than 700,000 yuan per year, far exceeding his budget (SID3). In 2011, as one of 24 young entrepreneurs who started and developed the enterprise, Mr. W participated in Shanghai Modern Service Industry Association and had exchanges with Shanghai leaders on the theme of "innovation and development of Shanghai modern service industry" (SIG3). In 2013, Shanghai Municipal Construction Commission set up a special fund to support the roof greening (not be less than 30%) of new buildings (SIG4).

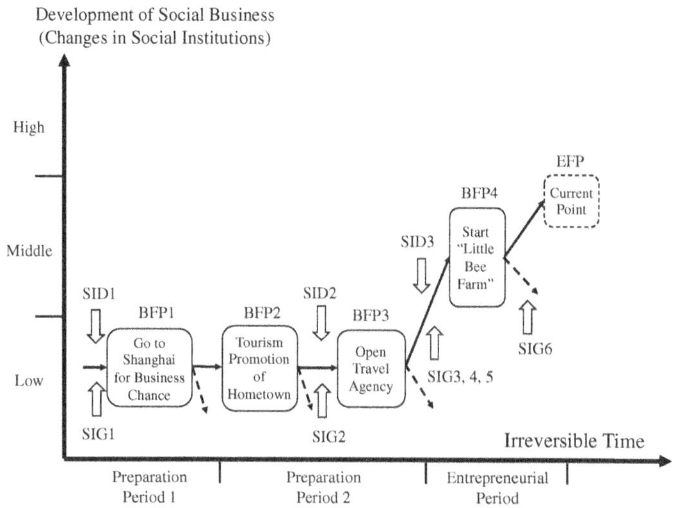

Fig. 6.7 TEM diagram of the analysis on Mr. W

Table 6.7 List of Mr. W's BFP, SIG and SID

	Contents
BFP1	Leaving for Shanghai to find business chance
2	Promoting tourism for his hometown
3	Opening travel agency "Datong travel"
4	Starting "little bee farm"
SIG1	Resigning a stable job
2	Spending 2 years to obtain the first fund
3	Active exchange with Shanghai leaders
4	The policy to support green roofs in Shanghai
5	Urban residents' yearn needs for clean and tidy villages, healthy food, rich local culture and farmers' hospitality
6	A global trend of food education
SID1	Against his parents' wishes
2	Unpopular as a tourist destination
3	Extremely high rent in the early days of the business

Mr. W is aware that urban residents yearn for clean and tidy villages, healthy food, rich local culture and farmers' hospitality (SIG5) on one side, and crave for better education and resources in megacities on the other. He started the first phase of construction of a "Little bee farm" (BFP4) with his partners. The Little bee farm is a high-quality center for ecology, play, scientific promotion, art and culture, and family love and is also a center of excellence for food education (food + agriculture + education) (SIG6).

6.4 Concluding Remarks

The above results of TEM analysis show that social entrepreneurship (SE) is affected by social institutions (SI), however it can also enhance changes in social institutions through the activities of social business (SB). It is also clarified that social capital (SC) is critically important to social entrepreneurship for the development of social business. However, with the development of social businesses that bring out the multi-functionality of agriculture, it is expected that the residents will have a deep understanding and active involvement in urban agriculture, if more stakeholders are involved.

Appendix 6.1: Sources of the TEM Analyses

See Tables 6.8 and 6.9.

Table 6.8 List of sources for the TEM analysis on Japan

Cases	Sources	Type
Mr. O	Ono (2018)	Book
	Hatakenbo HP <https://hatakenbo.org/>	Web
	Report: interim report on subsidies for housing and community building (2017) <https://www.hc-zaidan.or.jp/pdf/report/2501/05.pdf>	Web
	Report: activity report for housing and town planning NPO (2019) <https://www.hc-zaidan.or.jp/pdf/report/25/05.pdf>	Web
	Project Plan: Kunitachi City town planning project implementation plan utilizing agriculture and farmland (2012) <https://www.sangyo-rodo.metro.tokyo.jp/nourin/pdf/3machi/6kunitachi/kunitachi-keikaku.pdf>	Web
Mr. M	Miyaji (2009)	Book
	Miyaji and Saito (2017)	Article
	Miyaji (2019)	Book
Mr. N	Nishitsuji (2012)	Book
	MY FARM HP <https://myfarm.co.jp/>	Web

Table 6.9 List of sources for the TEM analysis on China

Cases	Sources (Web)
Mr. L	"Guāngmíng shè jiàoyù jiā—tā zài shànghǎi jiànle 90 duō gè shèqū huāyuán, gěi háizi jiā ménkǒu de zìrán jiàoyù" (2020) <https://news.tongji.edu.cn/info/1006/75103.htm>
	"Liúyuèlái: Zài shànghǎi jiànzào 2040 gè shèqū huāyuán" (2019) <http://www.zhhjw.org/a/qkzz/zzml/201906/rw/2019/0624/7175.html>
	"Zhòng cài yěyǒu dàxuéwèn? Tóngjì dà xué yǒu rén jiào nǐ zěnme zuò jǐngguān shūcài" (2020) <https://www.sohu.com/a/423691005_747096>
	"Tóngjì dàxué jǐngguān xué xì liúyuèlái lǎoshī tán shèqū nóng yuán" (2017) <http://www.chla.com.cn/htm/2017/0213/257690.html>
	"Sì yè cǎotáng 'ràng shèqū huāyuán' biàndìkāihuā" (2017) <https://www.whb.cn/zhuzhan/kandian/20170706/96821.html>
	"Shèqū huāyuán rén rén kě cānyù de 'wēi jījiàn'" (2020) <http://sh.xinhuanet.com/2020-11/19/c_139527961.htm>
	"Lèsè zàishēng lshèqū huāyuán shēngtài duīféi cāozuò zhǐnán" (2020) <https://www.sohu.com/a/367254847_260616>
Mr. Z	Tonys Farm HP <http://www.tonysfarm.com/site/about#level-1> <http://www.tonysfarm.com/site/strategy#level-1>
	"Tóngyàng shì zhǒng shūcài, yǒurén què piān yào tóu tā 7000 wàn!" (2018) <https://www.sohu.com/a/281559460_100141351>
	"Xīn nóngyè jīngyíng-duōlì nóngzhuāng: Bǎ cài mài gěi yǒu qián rén" (2012) <http://www.brand.zju.edu.cn/2012/0606/c57339a2294073/page.htm>
	"Zhāngtóngguì-tā yào tiàochū shànghǎi cānyǐn yè de guàiquān" (2004) <http://news.sina.com.cn/c/2004-12-31/20435390629.shtml>
	"Duōlì nóngzhuāng zhāngtóngguì: Zǔzǔbèibèi shì nóngmín, shūcài mài dào CBD" (2013) <http://finance.sina.com.cn/roll/20130415/052315145023.shtml>
	"Mài diào 30 jiā chuāncài guǎn, yīxīn gǎo nóngzhuāng dāng zhuāng zhǔ, yěshì zuìle" (2014) <https://www.toutiao.com/a3738620475/>
Mr. W	"Zài shànghǎi zū yīkuài dì, dài hái zǐ tǐyàn zhòng cài lèqù" (2019) <https://www.thepaper.cn/newsDetail_forward_3937238>
	"Shànghǎi xiǎo mìfēng nóngchǎng" (2020) <https://www.163.com/dy/article/F29DGV350518CELC.html>
	"Xiǎohuǒ fàngqì "tiě fànwǎn", chuàngbàn lǚxíngshè, yǒngyǒu jìn 40 jiā zhí yíng wǎngdiàn" (2018) <https://www.sohu.com/a/283233309_117373>
	"Ràng chuàngxīn chuàngyè jīngshén liútǎng zài chéngshì xiěyè zhōng" (2011) <http://news.sina.com.cn/o/2011-12-09/065023602521.shtml>
	"Wūdǐng lǜhuà zhèng chéngwéi chéngshì yī zhǒng cháoliú hé zhǐbiāo" (2015) <https://m.smokibacon.com/detail/2015112/224757.htm>

References

Miyaji Y (2009) Selling pigs in the Shonan wind. Kanki Publishing, Tokyo (in Japanese)
Miyaji Y (2019) Innovation by young generations in farm households. In: Kiminami A(ed) Entrepreneurship and innovation in Japanese agriculture. Springer, Singapore
Miyaji Y, Saito K (2017) Making Japanese agriculture smart, exciting, and profitable: Case studies of agricultural innovation by beginning young farmers. Japan J Farm Manage 55(1):12–22 (in Japanese)
Nishitsuji K (2012) MY FARM challenge from abandoned farmland: changing the society through business that connects farmers and people. Gakugei-Shuppan-sha, Kyoto (in Japanese)
Ono A (2018) Tokyo agricultural creators. Ikaros Publication Ltd., Tokyo (in Japanese)
Spulber DF (2014) Innovative entrepreneur. Cambridge University Press, New York

Chapter 7
Cognitive Innovation and Social Entrepreneurship in Urban Agriculture in Japan and China

Abstract Cognitive innovation consists of an endless cycle of exploration, exploitation and explanation; an exploratory process of probing boundaries, selecting which of the countless ideas to exploit and develop further, synthesizing the resulting new insights into explanations of how things are, generating new challenges and questions (Gummerum and Denham, Europe's J Psychol 10:586–588, 2014). The purpose of this chapter is to clarify the cognitive innovation of social entrepreneurs for creating shared value in urban agriculture in Japan and China through qualitative analysis based on the results of selected case studies in Chap. 6.

Keywords Cognitive innovation · Social entrepreneurship · Qualitative analysis · Learning capacity · Urban agriculture · Japan · China

7.1 Introduction

Cognitive innovation is an integrative concept that recognizes the fundamental links between the innovator, the innovation, the contextual challenges and consequential reverberations through society. Denham and Punt (2017) conceptualized cognitive innovation as "a recursive process in which an individual probe, its boundaries to seek out new knowledge, selects promising avenues for more extensive exploitation, and synthesizes what it learns within its growing body of knowledge, which includes knowledge of how to act in the world and how to interact with other individuals".

$$F_t + d_t,\ x_t + d_t \leftarrow F_t(F_t,\ x_t,\ s_t)$$

where F_t represents all internal (mental) processes, x_t is the set of ideas, facts, words, and so on that are known by the individual and can be exchanged with others, s_t are things in the world perceptually accessible to the individual and t is an index of time. Cognitive innovation refers to changes in F and x as the individual builds its knowledge, x, and processing capabilities and skills, represented by F. Cognitive innovation consists of an endless cycle of exploration, exploitation and explanation;

L. Kiminami et al., *Social Entrepreneurship, Social Business and the Multi-functionality of Urban Agriculture*, SpringerBriefs in Economics,
https://doi.org/10.1007/978-981-19-1762-2_7

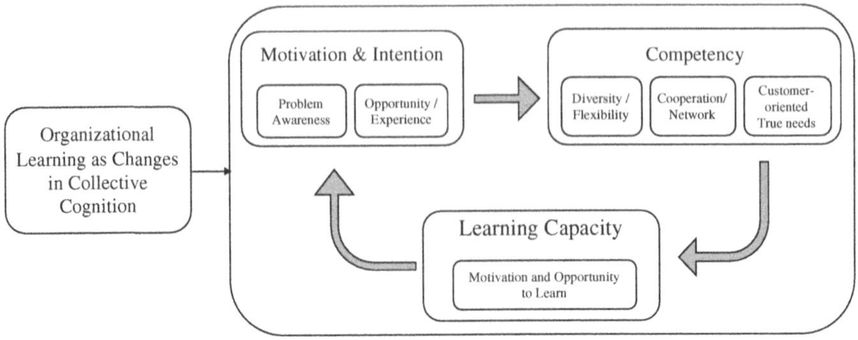

Fig. 7.1 Analytical framework

an exploratory process of probing boundaries, selecting which of the countless ideas to exploit and develop further, synthesizing the resulting new insights into explanations of how things are, generating new challenges and questions (Gummerum and Denham 2014).

In the following, we will clarify the cognitive innovation of social entrepreneurship in urban agriculture in Japan and China through qualitative analysis based on the framework as shown by Fig. 7.1. In setting the evaluation items for each element in this chapter, Yamamoto (2002), Hosoki and Kiminami (2021) were referred. We also added the learning ability that is assumed to be strongly related to the cognitive innovation of entrepreneurs. In other words, the development processes of social entrepreneurship are categorized through the analysis of each element (motivation/orientation, competency, learning ability, etc.) in phase I and phase II. Furthermore, we will superimpose and organize the aspects of business development as a social business that creates shared value. For information on each analysis target, we use the contents obtained from literature/materials and interview surveys as the same as collected in Chap. 6.

7.2 Analytical Results in Japan[1]

Figure 7.2 (see Appendix 7.1 for details) shows the changes of motives/orientations (problem awareness, opportunities/experiences) and competencies (diversity/flexibility, networks, customer orientations) in the first and second phases of social entrepreneurship with respect to the three cases in Japan. The results of evaluation on true needs and learning ability (learning opportunity/motivation) are given based on each element. Evaluation is performed on a 7-point scale (1: very weak, 2: weak, 3: slightly weak, 4: normal, 5: slightly strong, 6: strong, 7: very strong).

[1] See Chapter 6 for the details of the cases.

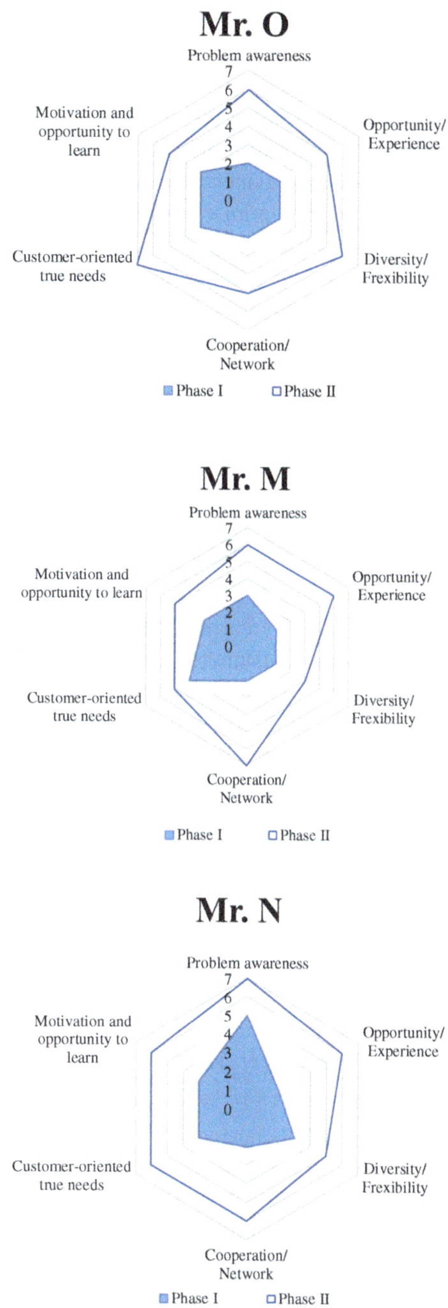

Fig. 7.2 Radar chart of entrepreneurs in urban agriculture of Japan

7.2.1 Case I: Mr. O

Mr. O was initially unaware of agricultural issues and attractiveness, didn't have high social entrepreneurship. In Phase I, each element was rather low, and the social entrepreneurship of him was gradually increased in the process of trial and error in search of social business in urban agriculture. In Phase II, he established a community farm, received administrative support, and clarified the concept through discussions with the government and local stakeholders. As a result, Mr. O is engaged in an innovative social business that provides space design and services for communities with diversity, openness, and comfort. He focuses on the design for various people to interact with each other, causing cognitive changes and value creation (shared value creation) through interaction between users and operators and among customers. The business is highly diverse, customer-oriented, and successfully managed to capture the true needs of users.

7.2.2 Case II: Mr. M

Mr. M attaches great importance to creating a face-to-face relationship between producers and consumers through the barbecue business as a motive for succeeding to the pig farming business at his parents' home. It can be said that he had relatively high customer orientation and true customer needs in Phase I, but his social entrepreneurial spirit itself was not high. In Phase II, Mr. M planned new business development while being restricted as a pig farming business located in the city, and rebuilt the internal cohesive social capital such as family, acquaintances / friends and local communities. In addition, Mr. M is transforming his entrepreneurial spirit into social entrepreneurship. He has built a bridging network inside and outside agriculture, as well as a platform for solving social problems, such as the establishment of a "farmer's child network."

7.2.3 Case III: Mr. N

Mr. N started his business with two clear missions: "Increase the number of people who can make their own food" and "Regenerate abandoned cultivated land" in phase I. In addition, he set up two courses to solve social problems: "Weekend Farming Preparation Course" and "Organic Professional Farmer Training Course" for supporting farming preparation and training organic farmers. He was strongly conscious and had a certain level of social entrepreneurship from the beginning. In Phase II, he started a project aimed at the reconstruction of agriculture in Tohoku area from the Great Earthquake of East Japan. Without fear of failure, he took it as an opportunity to think about various types of agriculture, while working in collaboration with local farmers. His awareness of the problem and each element of social entrepreneurship has been increased. In addition, he is planning various business developments such as distribution business, farmland rental service with sales destination, agriculture-related consulting business, and overseas business expansion.

7.3 Analytical Results in China[2]

Figure 7.3 (see Appendix 7.2 for details) shows the motives/orientations (problem awareness, opportunities/experiences) and competencies (diversity/flexibility, networks, customer orientations) in the first and second phases of social entrepreneurship with respect to the three cases in China. Since the social entrepreneurship of urban agriculture in China has only just developed in recent years, we will focus on the two changes between the preparatory period and the business development period. Specifically, the preparatory period II will be set as the social entrepreneurial preparatory period, the business development period will be set as the social entrepreneurial development period.

7.3.1 Case IV: Mr. L

Mr. L is demonstrating leadership in the establishment, operation and dissemination of community gardens in Shanghai through activities that are both experts (university teachers) and practitioners. In Phase I, he had a strong awareness of the problem of creating urban spaces to interact with nature and create opportunities for residents to interact with each other, and each element of social entrepreneurship had reached a high level as a whole. In Phase II, he strengthens the network and cooperation with local residents, governments, residents' committees, universities and companies, and promoting the participation of diverse residents inside and outside the region, enhances social entrepreneurship. Mr. L has a high evaluation of all factors as a whole, in particular, his problem awareness, cooperation/network, and customer orientation / true needs stand out.

7.3.2 Case V: Mr. Z

While developing the restaurant business, Mr. Z noticed a business opportunity in the agricultural field due to his interest in the problem of raw material procurement,

[2] Please see Chap. 6 for the details of the cases.

Fig. 7.3 Radar chart of entrepreneurs in urban agriculture of China

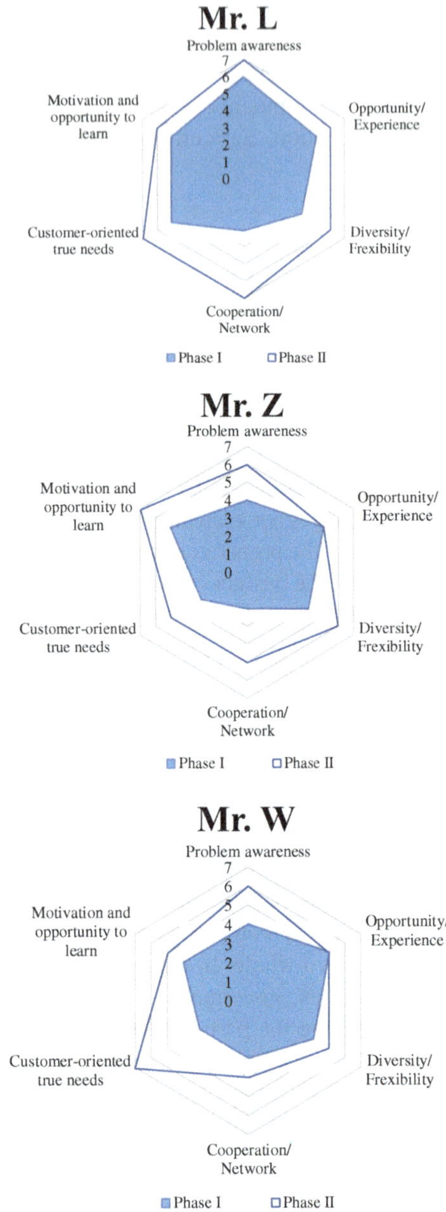

and started the organic farming business. He got the degree of EMBA to start the business and raised funds by selling the restaurant business, but the level of social entrepreneurship in Phase I was not high. However, when he started his organic farming business, the level of each element was rising rapidly. He also actively utilized agricultural and non-agricultural professional human resources to strengthen collaboration and networks with universities, researchers and farmers for business development. He presents a new business model for organic farming in China, which was not found in conventional farming, and features high motivation and learning opportunities.

7.3.3 Case VI: Mr. W

Mr. W is engaged in a rooftop farm business to utilize the city's rooftop space, improve the environment, and provide opportunities for agricultural experience and food education. In Phase I, as a person living in the city, the lifestyle and the way of raising children are reviewed and the awareness of the problem was clarified. As a member of a young entrepreneur, he had the opportunity to interact with leaders in the region, and he came up with the idea of a business from the perspectives of both consumers and entrepreneurs. In Phase II, a rooftop farm business that utilizes the rooftop space of a large-scale shopping mall as a place for agricultural experience/food education, art culture and science dissemination are established. The customer orientation and true needs are highly evaluated.

7.4 Pattern of Cognitive Innovation of Social Entrepreneurs

From the above results, the processes and the patterns of cognitive innovation of social entrepreneurs and the virtuous cycle of CSV can be organized into three main models as shown by Fig. 7.4. Each social entrepreneur has strong characteristics of their own and their social entrepreneurship was enhanced by cognitive innovation through trial and error and systematic learning.

7.4.1 Radical Cognitive Innovation (Mr. O and Mr. Z)

As for the two cases, each element was at a low level in Phase I but cognitive innovations were created in rapid progress along with the growth and expansion of the businesses. As a result, the factor evaluation in the entrepreneurial phase II is rising rapidly. In particular, Mr. O is highly evaluated for customer-oriented

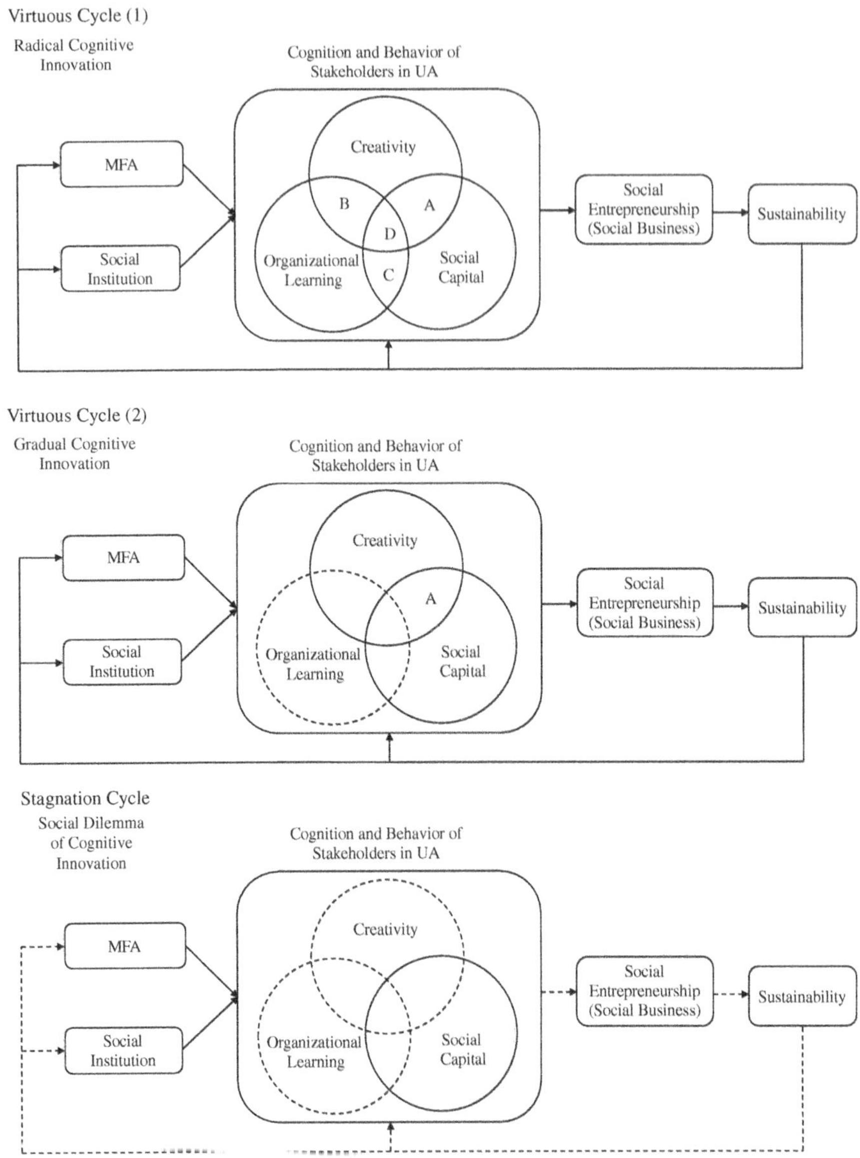

Fig. 7.4 Cognitive innovation and virtuous cycle for CSV

and true needs which were related to radical cognitive innovation. Mr. Z is highly evaluated for building and presenting an agricultural business model that didn't exist in China due to his high motivation and learning opportunities related to radical cognitive innovation. Therefore, the interactions of creativity, social capital, and organizational learning (the area of D in Fig. 7.4) among the stakeholders in urban agriculture promoted the social entrepreneurs' cognitive innovation.

7.4.2 Gradual Cognitive Innovation (Mr. N, Mr. M, Mr. L and Mr. W)

As for the four cases, each element was high to a certain extent even in the first phase of entrepreneurship, and the cognitive innovation is getting gradually as the business grows and expands. The elemental evaluations in the second phase of entrepreneurship are relatively well-balanced. However, more attention should be paid to the organizational learning for a higher level of cognitive innovation in these cases.

7.4.3 Stagnation: Social Dilemma of Cognitive Innovation[3]

If social institutional directions in the region are strong and social entrepreneurs are lack of open mind, they might be assimilated into the environment and become a part of it. Under the circumstances, they might not obtain social support from the region (Kiminami et al. 2020). Therefore, organizational learning involving the people surrounding seems to be necessary for the cognitive innovation of the entrepreneurs themselves.

Appendix 7.1: Evaluation of Elements of Social Entrepreneurial Activities in Japan

See Tables 7.1 and 7.2.

Appendix 7.2: Evaluation of Elements of Social Entrepreneurial Activities in China

See Tables 7.3 and 7.4.

[3] Please see Kiminami et al. (2020) for more details.

Table 7.1 Evaluation of elements in phase I (Japan)

	Mr. O	Mr. M	Mr. N
Problem awareness	Urban agriculture, where many people who do not know the production site of agriculture live, has important social significance *Weak = 2 points*	We sell delicious pigs directly to customers who can see their faces. A business model for pig farmers that is not pursuing scale *Slightly weak = 3 points*	Achieve two things: increase the number of people who make and eat by themselves, regenerate abandoned cultivated land, and revitalize Japanese agriculture *Somewhat strong = 5 points*
Opportunity/experience	Change jobs to an agricultural corporation and learn production technology *Weak = 2 points*	Emphasis on sending e-mail newsletters. I don't advertise. Sow seeds for the future by communicating the idea to the people you meet *Weak = 2 points*	Utilization of incubation office *Weak = 2 points*
Diversity· Flexibility	Although he focused on multipurpose space design, the concept is unclear *Weak = 2 points*	A business model that makes the best use of local resources (agriculture) *Weak = 2 points*	Sales to each site to find a farmer to cooperate with *Slightly weak = 3 points*
Cooperation/Network	Idea study and cooperation system with local residents *Weak = 2 points*	Cooperation with tourist farms Main customers are friends and acquaintances *Weak = 2 points*	Connection with young farmers in the area *Weak = 2 points*
Customer-oriented True needs	A farm where you can rent out fields and enjoy barbecue *Slightly weak = 3 points*	A face-to-face relationship with customers through barbecue. Develop your own sales channel *Normal = 4 points*	Providing opportunities for those who wish to experience farming *Slightly weak = 3 points*
Motivation and opportunity to learn	Learn from a series of challenges and failures. Learn through cooperation with people *Slightly weak = 3 points*	Learning to gain knowledge. Learning for entrepreneurship *Slightly weak = 3 points*	Humble learning to gain trust from sales failure experience *Slightly weak = 3 points*

Table 7.2 Evaluation of elements in phase II (Japan)

	Mr. O	Mr. M	Mr. N
Problem awareness	Creating a community space with diversity, openness, and comfort *Strong = 6 points*	Solve Japanese agricultural problems. 3 earn, if there is cool to-impress primary industry K to industry *Strong = 6 points*	I want to solve the problems of Japanese agriculture through agricultural service business that provides agricultural experience, farming support, and farmer training *Very strong = 7 points*
Opportunity/experience	"Vegetable making" DVD supervision, seminar/training instructor media appearance *Somewhat strong = 5 points*	Participated in various projects starting from the platform (NPO) created with friends *Strong = 6 points*	Cooperation in entrepreneurship/independence seminars, participation in the Great East Japan Earthquake reconstruction project, media appearance *Strong = 6 points*
Diversity·Flexibility	Accepting private lodgings and inbound tourists *Strong = 6 points*	Participate in a community where you can share your ideas and thoughts *Normal = 4 points*	Utilization measures and business development of abandoned cultivated land that meet the characteristics and needs of each region, overseas expansion *Somewhat strong = 5 points*
Cooperation/network	Collaboration with local residents, government, and university students *Somewhat strong = 5 points*	Social entrepreneurs, to this farmhouse, but les of organizational learning network. Companies, different industries, autonomous body, young farmers *Very strong = 7 points*	In cooperation with various organizations, agricultural experience, as well as human resource development business, research and development, flow through, until the sales business, perform a wide range of business development *Strong = 6 points*

(continued)

Table 7.2 (continued)

	Mr. O	Mr. M	Mr. N
Customer-oriented True needs	Entertainment-type agricultural services, child-rearing support business *Very strong = 7 points*	Create a platform for farmers to grow *Somewhat strong = 5 points*	The business is embodied by grasping various needs related to agriculture (consumers, local residents, producers, educational/research institutes) from multiple perspectives *Strong = 6 points*
Motivation and opportunity to learn	From "urban residents" to "typical consumers", meet agriculture and the community, and continue to grow with the process of becoming "local residents" with their families *Somewhat strong = 5 points*	The humility of being an "ordinary person without special abilities". Balance of knowledge and action. A challenging spirit that "you can't understand until you try (after absorbing knowledge)" *Somewhat strong = 5 points*	Prepared to continue learning even in different climates (Tohoku) and harsh conditions *Strong = 6 points*

Table 7.3 Evaluation of elements in phase I (China)

	Mr. L	Mr. Z	Mr. W
Problem awareness	Creating a space for urban agriculture that allows people to come into contact with nature and respond to the diluting of human relationships *Strong = 6 points*	Development of a restaurant business that offers Sichuan cuisine of hometown. Awareness of business opportunities in organic farming *Normal = 4 points*	Tourism promotion of hometown. Awareness of agricultural experience needs in the city. Disturbances in children's lives *Normal = 4 points*
Opportunity/experience	Architectural designer experience. Obtained a Ph.D. in city planning *Somewhat strong = 5 points*	Understanding the problem of raw material procurement in the food and beverage business. Obtained EMBA *Somewhat strong = 5 points*	City government policy on rooftop greening. Participation in the Shanghai Contemporary Service Industry Association. Interaction with local leaders as a young entrepreneur *Somewhat strong = 5 points*
Diversity·flexibility	Activities that are both experts and practitioners *Normal = 4 points*	Securing funds by selling a successful restaurant business *Normal = 4 points*	Reconsidering an urban lifestyle that comes from the sense of ownership as a city dweller *Normal = 4 points*
Cooperation/Network	Established an activity organization with friends *Slightly weak = 3 points*	Family opposition *Weak = 2 points*	Preparing to build a rooftop farm in collaboration with partners *Slightly weak = 3 points*
Customer-oriented true needs	Management with the participation of local residents *Somewhat strong = 5 points*	The needs of restaurant customers and consumers *Slightly weak = 3 points*	Customer needs for sightseeing trips *Slightly weak = 3 points*
Motivation and opportunity to learn	Experience of suggestions and rejections. Practical application of expertise *Somewhat strong = 5 points*	Acquisition of management knowledge and skills *Somewhat strong = 5 points*	Responding to changes in rooftop greening policies. Learning from overseas rooftop gardens *Normal = 4 points*

Table 7.4 Evaluation of elements in phase II (China)

	Mr. L	Mr. Z	Mr. W
Problem awareness	Construction and operation of urban farms with an emphasis on the principle of sustainability. Diffusion of CGs *Very strong = 7 points*	Business model of organic farming business for healthy and safe food supply *Strong = 6 points*	Business model for effective utilization of rooftop space, environmental improvement, agricultural experience and food education *Strong = 6 points*
Opportunity/experience	Involvement in promoting the diffusion of CGs in the city. Dissemination by media articles *Strong = 6 points*	Long-term investment projects such as soil improvement. Dissemination by media articles *Somewhat strong = 5 points*	Attention as a food education base in the world. Dissemination by media articles *Somewhat strong = 5 points*
Diversity·flexibility	Practice of a new business model under the constraints of the COVID-19 *Strong = 6 points*	Utilization of diverse professional human resources in agriculture and non-agriculture *Strong = 6 points*	Positioned as an opportunity to experience elements of art and culture *Somewhat strong = 5 points*
Cooperation/network	Collaboration with local residents, city governments, residents' committees, universities, companies, NPOs/NGOs *Very strong = 7 points*	Collaboration with universities, researchers, and local farmers. Loans from private equity institutions *Somewhat strong = 5 points*	Collaboration with governments, businesses and local residents *Normal = 4 points*
Customer-oriented true needs	Promoting the participation of diverse residents inside and outside the region *Very strong = 7 points*	Sales and agricultural experience business for the wealthy and middle class in urban areas. Membership business *Somewhat strong = 5 points*	Agricultural experience, food education. A place for art culture and science dissemination. Creating opportunities for parent–child interaction *Very strong = 7 points*
Motivation and opportunity to learn	Challenge to new business (seed and plant sharing) *Strong = 6 points*	Exploration and construction of management strategies not found in conventional agriculture *Very strong = 7 points*	Build a win–win relationship with a large shopping mall *Somewhat strong = 5 points*

References

Denham SL, Punt M (2017) Cognitive innovation: a view from the bridge. Leonardo 50(2):184–185

Gummerum M, Denham SL (2014) Cognitive innovation: from cell to society. Europe's J Psychol 10(4):586–588

Hosoki C, Kiminami L (2021) Issues on fostering social entrepreneurs in Japanese agriculture. Bulletins Faculty Agric Niigata Univ 73:29–38 (in Japanese)

Kiminami L, Furuzawa S, Kiminami A (2020) Social entrepreneurship and social business associated with multiple functions of urban agriculture in Japan. Asia-Pacific J Region Sci 4:521–552

Yamamoto C (2002) A study of social entrepreneur: Action model learning to innovation. Master thesis, Kochi University of Technology. http://www.kochi-tech.ac.jp/library/ron/2002/g5/M/1035024.pdf (in Japanese)

Chapter 8
Conclusion and Policy Implication

Abstract Policy implications drawn from the analytical results of this book suggested that social entrepreneurs need to carry out their businesses through balancing the economic and social aspects according to the actual conditions of the region while a place-based policy rather than a uniform support policy is also required. In addition, it was suggested that policy support needs to consider not only the business content and business plan of social entrepreneurs, but also whether entrepreneurs are willing to continuously engage in innovation in their relationships with society. Furthermore, the future research agendas for urban agriculture were also pointed out in this book.

Keywords Urban agriculture · Social business · Creative class · Creating shared value · Policy implication

8.1 Conclusions

This book clarified the mechanism of creating shared value through social business in urban agriculture using a mixed methods research. The relationships among creative class, social capital, multi-functionality of agriculture, and social business in Tokyo and Shanghai were clarified through quantitative analysis (SEM), and the cognitive innovation of social entrepreneurship in urban agriculture was grasped through qualitative analysis (TEM). In the following, we summarize the results obtained from each chapter.

Using bibliometric methods, the topics of previous research on urban agriculture were organized in Chap. 1. As a relatively new initiative, urban agriculture with the nature of social business was inactively positioned in the conventional research approach on urban economy and on agricultural and rural development. Therefore, the purpose of this book is to elucidate the mechanism by which social entrepreneurs solve regional issues through urban agriculture as a social business and contribute to sustainable urban development both in Japan and China, mainly focusing on the two megacities, Tokyo and Shanghai.

L. Kiminami et al., *Social Entrepreneurship, Social Business and the Multi-functionality of Urban Agriculture*, SpringerBriefs in Economics,
https://doi.org/10.1007/978-981-19-1762-2_8

A literature review was conducted in Chap. 2 focusing on the key (important) concepts of sustainable urban development: creativity and creative classes, social capital, the multi-functionality of urban agriculture, shared value creation, social entrepreneurship and social innovation. Based on the review results, it pointed out that the multi-functionality of urban agriculture should be reconsidered as a process of creating shared value through social business involved by creative classes with high level of social entrepreneurship and social capital. And empirical analysis on urban agriculture from the view of the creativity and sustainability of cities was called for.

A new conceptual framework was constructed in Chap. 3, by fusing the previous research in the fields of urban agriculture theory, creative class theory, social entrepreneurship theory, and organizational learning theory. Furthermore, a mixed methods research integrating both quantitative analysis and qualitative analysis was discussed and designed for hypothesis verification in the book. Specifically, structural equation modeling was introduced for clarifying the cognition and behavior of urban residents toward urban agriculture, and trajectory equifinality modeling was introduced for elucidating the mechanism of shared value creation.

Chapter 4 gave an overview of trends and policies in urban agriculture in Tokyo and Shanghai. It clarified that an institutional framework for urban agriculture had been put in place in Japan and the multi-functionality of urban agriculture was still at the initial stage of permeating among urban residents. Although the institutional framework for urban agriculture was underdeveloped in China, the shift of urban agriculture policy toward emphasizing the exertion of multiple functions is progressing. However, in the two countries, few studies have clarified the development mechanism of urban agriculture as a social business involved by creative classes who are urban residents and entrepreneurs with high level of social entrepreneurship and social capital.

It was clarified in Chap. 5 that people with creative thinking and a high level of social capital evaluated the multi-functionality of urban agriculture through a quantitative analysis (SEM) on the results of a questionnaire survey of urban residents in Tokyo and Shanghai. Therefore, it is supposed that some social problems such as lack of fresh and safe agricultural products, weakening of local communities, and deterioration of natural environment etc. can be solved to a certain extent in megacities by involving creative class in urban agriculture.

Furthermore, it was clarified in Chap. 6 through a qualitative analysis on the three selected cases of urban agriculture in each city that although social entrepreneurship was influenced by social institutions, social business can cause transformation in social institutions. In particular, it became clear that social capital plays an important role in these changes. The mechanism of shared value creation in urban agriculture was clarified in Chap. 7 by focusing on the cognitive innovation of entrepreneurs from three aspects of motivation/orientation, competency, and learning ability of urban agriculture entrepreneurs, based on the results of the case analysis in Chap. 6. And the patterns of cognitive innovation of entrepreneurship in urban agriculture were organized into three (radical cognitive innovation, gradual cognitive innovation, and social dilemma of cognitive innovation), in which organizational learning played an important role.

Therefore, the path to sustainability in urban agriculture linking with social entrepreneurship, social business and the multi-functionality of agriculture is shown by Fig. 8.1, in which horizontal axis represents economic value, and vertical axis represents social and environmental values. The curves in the figure represent the levels of value creation achievable under the resources and technology of urban agriculture at a certain point of time (close to the concept of a productivity curve). In addition, Am and Bm are the minimum levels required for agriculture industry to maintain sustainability, and the areas above these levels are designated as sustainability areas. Traditionally, urban agriculture has been regarded as unsatisfactory in terms of both economic and social and environmental values under the pressure of urbanization (E_1). However, the frontier is expanding through the development of business models that take advantage of being close to consumers and the utilization of advanced technology ($A_1B_1 \rightarrow A_2B_2$). This can be explained as the development of urban agriculture that takes advantage of the externalities of agglomeration in cities. Although the number of urban agricultural management increased, some of them had only reached the minimum level by increasing economic efficiency (E_2) or social / environmental value ($E_{2'}$). There were a few cases of urban farming that met sustainable level. On the other hand, as shown in the analytical results for Tokyo and Shanghai, creative classes are actively developing urban agriculture as a social business while solving local social problems. As a result, the frontier is expanding ($A_2B_2 \rightarrow A_3B_3$). In addition, the cases analyzed in this book are thought to have reached the sustainability area (E_3*) through Creating Shared Value (CSV) with stakeholders. On the other hand, it is considered that if social entrepreneurs fall into the dilemma of cognitive innovation may not reach the level of sustainability (E_3, E_3').

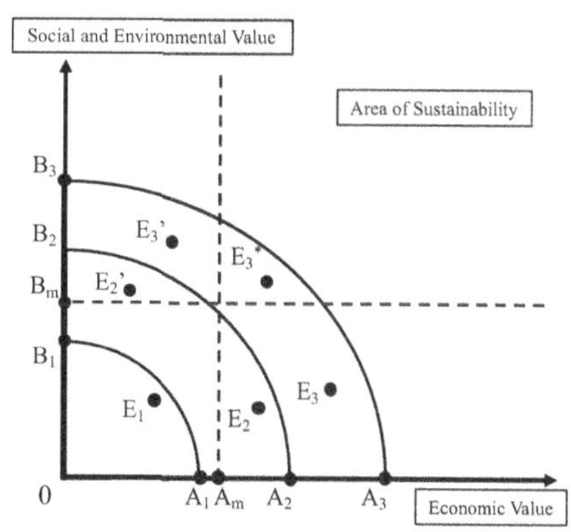

Fig. 8.1 Path to sustainability in urban agriculture. *Source* Revised from Kiminami and Kiminami (2017, Fig. 1)

8.2 Policy Implications and Future Research Agendas

Although, selected cases of social entrepreneurship in the book have reached sustainability in business through creating shared value in urban agriculture, there are still limited successful cases of social entrepreneurs working on solving social problems through urban agriculture both in Tokyo and Shanghai. In the process of business development, even if they succeed temporarily, there are cases where it cannot be sustained. The following two possible reasons are considered for these situations. The first is that cognitive innovation relies on the structure of the society that social entrepreneurs face and their relationship with the society. It is important for social entrepreneurs to involve stakeholders in cognitive innovation while they are influenced by social direction and social guidance in the process of business development. In other words, the level of economic and sociality and their balance required for CSV success are not uniform in all regions. Therefore, social entrepreneurs are required to carry out their businesses while balancing economic and social aspects according to the actual conditions of the region. In addition, policy makers are also required to have a place-based policy rather than a uniform support policy. The second is that cognitive innovation is a recursive process. Even if an entrepreneur has creativity, a high level of social capital, and learning ability at some point, there is no guarantee that he or she will be able to solve social problems. Moreover, even if a solution to a social problem is temporarily realized, it cannot always be sustained. Therefore, policy support needs to consider not only the business content and business plan of social entrepreneurs, but also whether entrepreneurs are willing to continuously engage in innovation in their relationships with society.[1]

The following issues can be pointed out as the agenda of future research. The first is the expansion of analysis targets and the refinement of analysis methods. In particular, it is necessary to combine quantitative and qualitative analysis in the application of mixed methods approach, and to theorize and simulate social innovation through urban agriculture. In addition, it is important to clarify the mechanism of virtuous cycle of three relationships (fostering creativity, accumulating social capital, and organizational learning) through quantitative research. Finally, analysis targeting farmers and entrepreneurs, and analysis of the actual situation of cooperation among various stakeholders in each case are necessary.

[1] This is similar to the difference between the two approaches in assessing a company's environmental protection activities. One is to evaluate by achieving the environmental standard value, and the other is to evaluate by the presence or absence of a continuous PDCA cycle to reduce the environmental load.

Reference

Kiminami L, Kiminami A (2017) Rural and agriculture development in regional science. In: Shibusawa H, Sakurai K, Mizunoya T, Uchida S (eds) Socioeconomic environmental policies and evaluations in regional science: essays in honor of Yoshiro Higano, New frontiers in regional science; Asian perspectives 24. Springer, Singapore, pp 537–555